T0291848

CAMBRIDGE LIBRARY COLLECTION

Books of enduring scholarly value

Physical Sciences

From ancient times, humans have tried to understand the workings of the world around them. The roots of modern physical science go back to the very earliest mechanical devices such as levers and rollers, the mixing of paints and dyes, and the importance of the heavenly bodies in early religious observance and navigation. The physical sciences as we know them today began to emerge as independent academic subjects during the early modern period, in the work of Newton and other 'natural philosophers', and numerous sub-disciplines developed during the centuries that followed. This part of the Cambridge Library Collection is devoted to landmark publications in this area which will be of interest to historians of science concerned with individual scientists, particular discoveries, and advances in scientific method, or with the establishment and development of scientific institutions around the world.

The Life and Letters of Faraday

Michael Faraday (1791-1867) made foundational contributions in the fields of physics and chemistry, notably in relation to electricity. One of the greatest scientists of his day, Faraday held the position of Fullerian Professor of Chemistry at the Royal Institution of Great Britain for over thirty years. Not long after his death, his friend Henry Bence Jones attempted 'to join together his words, and to form them into a picture of his life which may be almost looked upon as an autobiography.' Jones' compilation of Faraday's manuscripts, letters, notebooks, and other writings resulted in this *Life and Letters* (1870) which remains an important resource for learning more about one of the most influential scientific experimentalists of the nineteenth century. Volume 2 (1831–1867) describes his research on electricity and electromagnetism, his work as a scientific adviser to the government and industry and his service to education.

Cambridge University Press has long been a pioneer in the reissuing of out-of-print titles from its own backlist, producing digital reprints of books that are still sought after by scholars and students but could not be reprinted economically using traditional technology. The Cambridge Library Collection extends this activity to a wider range of books which are still of importance to researchers and professionals, either for the source material they contain, or as landmarks in the history of their academic discipline.

Drawing from the world-renowned collections in the Cambridge University Library, and guided by the advice of experts in each subject area, Cambridge University Press is using state-of-the-art scanning machines in its own Printing House to capture the content of each book selected for inclusion. The files are processed to give a consistently clear, crisp image, and the books finished to the high quality standard for which the Press is recognised around the world. The latest print-on-demand technology ensures that the books will remain available indefinitely, and that orders for single or multiple copies can quickly be supplied.

The Cambridge Library Collection will bring back to life books of enduring scholarly value (including out-of-copyright works originally issued by other publishers) across a wide range of disciplines in the humanities and social sciences and in science and technology.

The Life
and Letters of
Faraday

VOLUME 2

BENCE JONES

CAMBRIDGE UNIVERSITY PRESS

Cambridge, New York, Melbourne, Madrid, Cape Town, Singapore,
São Paolo, Delhi, Dubai, Tokyo

Published in the United States of America by Cambridge University Press, New York

www.cambridge.org
Information on this title: www.cambridge.org/9781108014601

© in this compilation Cambridge University Press 2010

This edition first published 1870
This digitally printed version 2010

ISBN 978-1-108-01460-1 Paperback

FARADAY'S LIFE AND LETTERS.

VOL. II.

FARADAY'S LABORATORY AT THE ROYAL INSTITUTION.

THE

LIFE AND LETTERS

OF

F A R A D A Y.

BY

DR. BENCE JONES,

SECRETARY OF THE ROYAL INSTITUTION.

IN TWO VOLUMES.

VOL. II.

SECOND EDITION, REVISED.

LONDON:

LONGMANS, GREEN, AND CO.

1870.

CONTENTS

OF

THE SECOND VOLUME.

———◆◆———

LIST OF ILLUSTRATIONS.

FARADAY'S STUDY AT THE ROYAL INSTITUTION.

LIFE OF FARADAY.

CHAPTER I.

FIRST PERIOD OF ELECTRICAL RESEARCH — DISCOVERY OF MAG-
NETO-ELECTRICITY — INDUCTION CURRENTS AND DEFINITE
ELECTRICAL DECOMPOSITION.

IT will be the object of this chapter first to describe
the great scientific work which Faraday did at this
period ; secondly, by means of his titles and the letters
which he received, to show the reputation he obtained
in consequence of his discoveries ; and thirdly, as far as
possible by means of his own letters, to give a picture
of the character which he made and kept during the time
of his great success.

I.

On August 29, 1831, Faraday began his ' Electrical
Researches.'

In December 1824 he believed with all his energy
that as voltaic electricity powerfully affects a magnet,
so the magnet ought to exert a reaction upon the
electric current. Guided by this idea, he made an experiment, of which one part (the passage of a magnet
through a metallic helix connected with a galvanometer),
if separated from the rest of the experiment, would then
have made the great discovery of magneto-electricity.

This experiment he published in the 'Quarterly Journal of Science,' July 1825, p. 338.

In November 1825, also, he had failed to discover voltaic induction. He passed a current through one wire, which was lying close to another wire, which communicated with a galvanometer, and found ' no result.' The momentary existence of the phenomena of induction then escaped him.

Again, December 2, 1825, and April 22, 1828, he made experiments which gave ' no result.' These experiments were not published.

The good time was now come. The first paragraph in the laboratory note-book is, ' Experiments on the production of electricity from magnetism.' His first experiment, detailed in the second paragraph, records the discovery by which he will be for ever known.

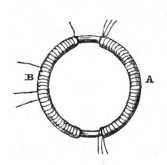

'I have had an iron ring made (soft iron), iron round and $\frac{7}{8}$ths of an inch thick, and ring six inches in external diameter. Wound many coils of copper round, one half of the coils being separated by twine and calico; there were three lengths of wire, each about twenty-four feet long, and they could be connected as one length, or used as separate lengths. By trials with a trough each was insulated from the other. Will call this side of the ring A. On the other side, but separated by an interval, was wound wire in two pieces, together amounting to about sixty feet in length, the direction being as with the former coils. This side call B.

'Charged a battery of ten pairs of plates four inches square. Made the coil on B side one coil, and connected its extremities by a copper wire passing to a distance, and just over a magnetic needle (three feet from wire ring), then connected the ends of one of the pieces on A side with battery : immediately a sensible effect on needle. It oscillated and settled at last in original position. On breaking connection of A side with battery, again a disturbance of the needle.'

In the 17th paragraph, written on the 30th of August, he says, ' May not these transient effects be connected with causes of difference between power of metals at rest and in motion in Arago's experiments?'

After this he prepared fresh apparatus. Writing to his friend R. Phillips, September 23, he says, ' I am busy just now again on electro-magnetism, and think I have got hold of a good thing, but can't say. It may be a weed instead of a fish that, after all my labour, I may at last pull up.'

September 24 was the third day of his experiments. He began paragraph 21 by trying to find the effect of one helix of wire, carrying the voltaic current of ten pairs of plates, upon another wire connected with a galvanometer. ' No induction sensible.' Paragraph 22. Longer and different metallic helices showed no effect, so he gave up those experiments for that day, and tried the effect of bar magnets instead of the ring magnet he had used on the first day.

In paragraph 33 he says, ' An iron cylinder had a helix wound on it. The ends of the wires of the helix were connected with the indicating helix at a distance by copper wire. Then the iron placed between the poles of bar magnets as in accompanying figure. Every time the magnetic contact at N or S was made or broken, there was magnetic motion at the indicating helix—the effect being, as in former cases, not permanent, but a mere momentary push or pull. But if the electric communication (i.e. by the copper wire) was broken, then the disjunction and  contacts produced no effect whatever. Hence here distinct conversion of magnetism into electricity.'

The fourth day of work was October 1. Paragraphs 36, 37, and 38 describe the discovery of induced voltaic currents.

' 36. A battery of ten troughs, each of ten pairs of plates four inches square, charged with good mixture of sulphuric and nitric acid, and the following experiments made with it in the following order.

' 37. One of the coils (of a helix of copper wire 203 feet long) was connected with the flat helix, and the other (coil of same length round same block of wood) with the poles of the battery (it having been found that there was no metallic contact between the two); the magnetic needle at the indicating flat helix was affected, but so little as to be hardly sensible.

' 38. In place of the indicating helix, our galvanometer was used, and then a sudden jerk was perceived when the battery communication was *made* and *broken*, but it was so slight as to be scarcely visible. It was one way when made, the other when broken, and the needle took up its natural position at intermediate times.

' Hence there is an inducing effect without the presence of iron, but it is either very weak or else so

1831.

Æt. 40.

sudden as not to have time to move the needle. I rather suspect it is the latter.'

The fifth day of experiment was October 17. Paragraph 57 describes the discovery of the production of electricity by the approximation of a magnet to a wire.

'A cylindrical bar magnet three-quarters of an inch in diameter, and eight inches and a half in length, had one end just inserted into the end of the helix cylinder (220 feet long) ; then it was quickly thrust in the whole length, and the *galvanometer* needle moved; then pulled out, and again the *needle moved*, but in the opposite direction. This effect was repeated every time the magnet was put in or out, and therefore a wave of electricity was so produced from *mere approximation of a magnet*, and not from its formation *in situ.'*

The ninth day of his experiments was October 28, and this day he 'made a copper disc turn round between the poles of the great horse-shoe magnet of the Royal Society. The axis and edge of the disc were connected with a galvanometer. The needle moved as the disc turned.' The next day that he made experiments, November 4, he found 'that a copper wire one-eighth of an inch drawn between the poles and conductors produced the effect.' In his paper, when describing the experiment, he speaks of the metal cutting the magnetic curves, and in a note to his paper he says, 'By magnetic curves I mean lines of magnetic forces which would be depicted by iron filings.'

This is the germ of those 'lines of force' which rose up in the mind of Faraday into 'physical' and almost tangible matter. The influence which they had upon his thoughts and experiments will be seen from this time up to the date of the last researches which he sent to the Royal Society in 1860.

In ten days of experiment these splendid results were obtained. He collected the facts into the first series of 'Experimental Researches in Electricity.' It was read, November 24th, at the Royal Society. Then he went to Brighton, and from thence, November 29th, he sends an abstract of this paper in a letter to his friend R. Phillips.

FARADAY TO PHILLIPS.

'Brighton: November 29, 1831.

'Dear Phillips,—For once in my life I am able to sit down and write to you without feeling that my time is so little that my letter must of necessity be a short one ; and accordingly I have taken an extra large sheet of paper, intending to fill it with news. And yet, as to news, I have none, for I withdraw more and more from society, and all I have to say is about myself.

'But how are you getting on ? Are you comfortable ? And how does Mrs. Phillips do ; and the girls ? Bad correspondent as I am, I think you owe me a letter ; and as in the course of half an hour you will be doubly in my debt, pray write us, and let us know all about you. Mrs. Faraday wishes me not to forget to put her kind remembrances to you and Mrs. Phillips in my letter.

'To-morrow is St. Andrew's day,[1] but we shall be here until Thursday. I have made arrangements to be

---

[1] The day of election of the new Council of the Royal Society.

out of the Council, and care little for the rest, although I should, as a matter of curiosity, have liked to see the Duke in the chair on such an occasion.

' We are here to refresh. I have been working and writing a paper that always knocks me up in health, but now I feel well again, and able to pursue my subject ; and now I will tell you what it is about. The title will be, I think, " Experimental Researches in Electricity : "—I. On the Induction of Electric Currents ; II. On the Evolution of Electricity from Magnetism ; III. On a new Electrical Condition of Matter ; IV. On Arago's Magnetic Phenomena. There is a bill of fare for you ; and, what is more, I hope it will not disappoint you. Now the pith of all this I must give you very briefly ; the demonstrations you shall have in the paper when printed.

' I. When an electric current is passed through one of two parallel wires, it causes at first a current in the same direction through the other, but this induced current does not last a moment, notwithstanding the inducing current (from the voltaic battery) is continued ; all seems unchanged, except that the principal current continues its course. But when the current is stopped, then a return current occurs in the wire under induction, of about the same intensity and momentary duration, but in the opposite direction to that first formed Electricity in currents therefore exerts an inductive action like ordinary electricity, but subject to peculiar laws. The effects are a current in the same direction when the induction is established ; a reverse current when the induction ceases, and a *peculiar state* in the interim. Common electricity probably does the same thing ; but as it is at present impossible to separate the beginning and the end of a spark or discharge from each other, all the effects are simultaneous and neutralise each other.

'II. Then I found that magnets would induce just like voltaic currents, and by bringing helices and wires and jackets up to the poles of magnets, electrical currents were produced in them ; these currents being able to deflect the galvanometer, or to make, by means of the helix, magnetic needles, or in one case even to give a spark. Hence the evolution of *electricity from magnetism*. The currents were not permanent. They ceased the moment the wires ceased to approach the magnet, because the new and apparently quiescent state was assumed, just as in the case of the induction of currents. But when the magnet was removed, and its induction therefore ceased, the return currents appeared as before. These two kinds of induction I have distinguished by the terms *volta-electric* and *magneto-electric* induction. Their identity of action and results is, I think, a very powerful proof of M. Ampère's theory of magnetism.

'III. The new electrical condition which intervenes by induction between the beginning and end of the inducing current gives rise to some very curious results. It explains why chemical action or other results of electricity have never been as yet obtained in trials with the magnet. In fact, the currents have no sensible duration. I believe it will explain perfectly the *transference of elements* between the poles of the pile in decomposition. But this part of the subject I have reserved until the present experiments are completed ; and it is so analogous, in some of its effects, to those of Ritter's secondary piles, De la Rive and Van Beek's

peculiar properties of the poles of a voltaic pile, that I 1831.
should not wonder if they all proved ultimately to Æt. 40.
depend on this state. The condition of matter I have
dignified by the term *Electrotonic*, THE ELECTROTONIC
STATE. What do you think of that? Am I not a bold
man, ignorant as I am, to coin words? but I have consulted the scholars. And now for IV.

' IV. The new state has enabled me to make out and
explain all Arago's phenomena of the rotating magnet or
copper plate, I believe, perfectly ; but as great names
are concerned (Arago, Babbage, Herschel, &c.), and as
I have to differ from them, I have spoken with that
modesty which you so well know you and I and John
Frost [1] have in common, and for which the world so
justly commends us. I am even half afraid to tell you
what it is. You will think I am hoaxing you, or else in
your compassion you may conclude I am deceiving
myself. However, you need do neither, but had better
laugh, as I did most heartily when I found that it was
neither attraction nor repulsion, but just one of my
old rotations in a new form. I cannot explain to you
all the actions, which are very curious ; but in consequence of the electrotonic state being assumed and
lost as the parts of the plate whirl under the pole, and
in consequence of magneto-electric induction, currents
of electricity are formed in the direction of the radii ;
continuing, for simple reasons, as long as the motion
continues, but ceasing when that ceases. Hence the
wonder is explained that the metal has powers on the
magnet when moving, but not when at rest. Hence is
also explained the effect which Arago observed, and

---

[1] A pushing acquaintance, who, without claim of any kind, got himself
presented at Court.

which made him contradict Babbage and Herschel, and say the power was repulsive ; but, as a whole, it is really tangential. It is quite comfortable to me to find that experiment need not quail before mathematics, but is quite competent to rival it in discovery ; and I am amazed to find that what the high mathematicians have announced as the *essential condition* to the rotation— namely, that *time is required*—has so little foundation, that if the time could by possibility be anticipated instead of being required—i.e. if the currents could be formed *before* the magnet came over the place instead of *after*—the effect would equally ensue. Adieu, dear Phillips.

' Excuse this egotistical letter from yours very faithfully,

' M. FARADAY.'

On December 5, 1831, Faraday was again at work in continuation of his researches.

For three days he at first occupied himself with more precise observations on the directions of the induced currents ; and on December 14, paragraph 217, he ' tried the effects of terrestrial magnetism in evolving electricity. Obtained beautiful results.'

' The helix had the soft iron cylinder (freed from magnetism by a full red heat and cooling slowly) put into it, and it was then connected with the galvanometer by wires eight feet long ; then inverted the bar and helix, and immediately the needle moved ; inverted it again, the needle moved back ; and, by repeating the motion with the oscillations of the needle, made the latter vibrate 180°, or more.'

The same day he ' made Arago's experiment with

the earth magnet, only no magnet used, but the plate
put horizontal and rotated.   The effect at the needle
was slight, but very distinct.'

1831.

Æt. 40.

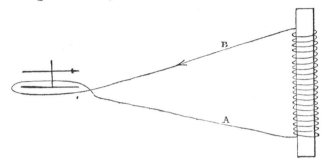

Paragraph 227.—'Hence Arago's plate, a new electrical machine.'   On four other days of December he
experimented on terrestrial magneto-electric induction,
and on voltaic electric induction.

In January he experimented on the currents produced
by the earth's rotation—on the 10th at the lake in the
gardens of Kensington Palace, and on the 12th and
13th at Waterloo Bridge.

February 8. Paragraph 423 says, 'This evening,
at Woolwich, experimented with magnet,[1] and for the
first time got the magnetic
spark myself.   Connected
ends of a helix into two
general ends, and then
crossed the wires in such a
way that a blow at $a\,b$
would open them a little.

Then bringing $a\,b$ against the poles of a magnet, the
ends were disjoined, and bright sparks resulted.

[1] The great magnet of the Royal Society was at this time at Mr. Christie's.

February 9th.—'At home succeeded beautifully with
Mr. Daniell's magnet. Amalgamation of wires very
needful. This is a natural loadstone, and perhaps the
first used for the spark.'

These, and many other experiments which he made
in December and January, he sent to the Royal Society,
and his paper on terrestrial magneto-electric induction,
and on the force and direction of magneto-electric
induction generally, was read as the Bakerian lecture,
January 12, 1832.

Professor Tyndall gives the following excellent condensation of this second paper :—

' He placed a bar of iron in a coil of wire, and lifting
the bar into the direction of the dipping needle, he excited by this action a current in the coil. On reversing
the bar, a current in the opposite direction rushed
through the wire. The same effect was produced, when,
on holding the helix in the line of dip, a bar of iron was
thrust into it. Here, however, the earth acted on the
coil through the intermediation of the bar of iron. He
abandoned the bar, and simply set a copper-plate spin-
ning in a horizontal plane ; he knew that the earth's
lines of magnetic force then crossed the plate at an
angle of about 70°. When the plate spun round, the
lines of force were intersected and induced currents
generated, which produced their proper effect when
carried from the plate to the galvanometer. " When the
plate was in the magnetic meridian, or in any other
plane coinciding with the magnetic dip, then its rotation
produced no effect upon the galvanometer."
' At the suggestion of a mind fruitful in suggestions of

a profound and philosophic character—I mean that of Sir John Herschel—Mr. Barlow, of Woolwich, had experimented with a rotating iron shell. Mr. Christie had also performed an elaborate series of experiments on a rotating iron disc. Both of them had found that when in rotation the body exercised a peculiar action upon the magnetic needle, deflecting it in a manner which was not observed during quiescence; but neither of them was aware at the time of the agent which produced this extraordinary deflection. They ascribed it to some change in the magnetism of the iron shell and disc.

' But Faraday at once saw that his induced currents must come into play here, and he immediately obtained them from an iron disc. With a hollow brass ball, moreover, he produced the effects obtained by Mr. Barlow. Iron was in no way necessary : the only condition of success was that the rotating body should be of a character to admit of the formation of currents in its substance : it must, in other words, be a conductor of electricity. The higher the conducting power, the more copious were the currents. He now passes from his little brass globe to the globe of the earth. He plays like a magician with the earth's magnetism. He sees the invisible lines along which its magnetic action is exerted, and, sweeping his wand across these lines, he evokes this new power. Placing a simple loop of wire round a magnetic needle, he bends its upper portion to the west : the north pole of the needle immediately swerves to the east : he bends his loop to the east, and the north pole moves to the west. Suspending a common bar magnet in a vertical position, he causes it to spin round its own axis. Its pole being connected with one end of a galvanometer wire, and its equator with the other end, electricity rushes round the galvanometer from the rotating magnet. He remarks upon the "*singular independence*" of the magnetism and the body of the magnet which carries it. The steel behaves as if it were isolated from its own magnetism.

'And then his thoughts suddenly widen, and he asks himself whether the rotating earth does not generate induced currents as it turns round its axis from west to east. In his experiment with the twirling magnet the galvanometer wire remained at rest; one portion of the circuit was in motion *relatively* to *another portion*. But in the case of the twirling planet the galvanometer wire would necessarily be carried along with the earth; there would be no relative motion. What must be the consequence? Take the case of a telegraph wire with its two terminal plates dipped into the earth, and suppose the wire to lie in the magnetic meridian. The ground underneath the wire is influenced, like the wire itself, by the earth's rotation; if a current from south to north be generated in the wire, a similar current from south to north would be generated in the earth under the wire: these currents would run against the same terminal plate, and thus neutralise each other.

'This inference appears inevitable, but his profound vision perceived its possible invalidity. He saw that it was at least possible that the difference of conducting power between the earth and the wire might give one an advantage over the other, and that thus a residual or differential current might be obtained. He combined wires of different materials, and caused them to act in opposition to each other, but found the combination ineffectual. The more copious flow in the better conductor was exactly counterbalanced by the resistance of the worst. Still, though experiment was thus emphatic,

he would clear his mind of all discomfort by operating on the earth itself. He went to the round lake near Kensington Palace, and stretched 480 feet of copper wire, north and south, over the lake, causing plates soldered to the wire at its ends to dip into the water. The copper wire was severed at the middle, and the severed ends connected with a galvanometer. No effect whatever was observed. But though quiescent water gave no effect, moving water might. He therefore worked at Waterloo Bridge for three days, during the ebb and flow of the tide, but without any satisfactory result. Still he urges, "Theoretically it seems a necessary consequence, that where water is flowing there electric currents should be formed. If a line be imagined passing from Dover to Calais through the sea and returning through the land, beneath the water, to Dover, it traces out a circuit of conducting matter, one part of which, when the water moves up or down the channel, is cutting the magnetic curves of the earth, whilst the other is relatively at rest. . . . There is every reason to believe that currents do run in the general direction of the circuit described, either one way or the other, according as the passage of the waters is up or down the Channel." This was written before the submarine cable was thought of, and he once informed me that actual observation upon that cable had been found to be in accordance with his theoretic deduction.'

In addition to this noble work, which placed him among the first philosophers, other work was done. Dr. Tyndall says :—

'In 1831 Faraday published a paper "On a peculiar class of Optical Deceptions," to which I believe the beautiful optical toy called the chromatrope owes its origin. In the same year he published a paper in the " Philosophical Transactions," " On Vibrating Surfaces," in which he solved an acoustical problem which, though of extreme simplicity *when solved*, appears to have baffled many eminent men. The problem was to account for the fact that light bodies, such as the seed of lycopodium, collected at the vibrating parts of sounding plates, while sand ran to the nodal lines. Faraday showed that the light bodies were entangled in the little whirlwinds formed in the air over the places of vibration, and through which the heavier sand was readily projected.'

He had also a paper in the 'Royal Institution Journal,' ' On the Limits of Vaporisation.' After Easter he gave four afternoon lectures on optical deceptions, lithography, flowing sand, and caoutchouc ; and during the season he took five Friday evenings for his discourses. One of these was on oxalamide, lately discovered by M. Dumas. His notes run thus :—' Is an artificial substance, yet approaching to organic matter. Wood distilled, acid—isinglass distilled, alkali ; hence the interest. Not one to think that by battery and retort we may make mind and body, but still feel free to observe effects, as far as we can trace them.' The other lectures were on a peculiar class of optical deceptions ; on light and phosphorescence ; on Trevelyan's recent experiments, on the production of sound during the conduction of heat ; and on the arrangements assumed by particles upon vibrating elastic surfaces.

II.

The only title he received this year was that of honorary member of the Imperial Academy of Sciences, St. Petersburg.

III.

The sensitiveness of his character, when a misinterpretation was given to his words, appears in a letter which he wrote to M. Gay-Lussac, regarding the first and second series of ' Experimental Researches.' The circumstances were these :—

Three weeks after Faraday's first paper was read at the Royal Society, he wrote a 'short' and 'hasty' and 'unfortunate' letter to M. Hachette, who communicated it a week afterwards to the Academy of Sciences, Paris, on December 26. Two days afterwards, the account was printed in 'Le Temps.' M. Nobili saw this, and, with M. Antinori, he immediately 'considered the subject was given to the philosophical world for general pursuit.' Their results were dated January 31, 1832, and they were published in the 'Antologia,' which was dated November 1831. Of this Faraday said, 'The circumstance of back date has caused many here who have heard of Nobili's experiments by report only, to imagine his results were anterior to, instead of being dependent upon mine.'

Faraday's second paper was read, January 12, 1832. Nobili and Antinori wrote a second memoir, dated March 1832. In the 'Philosophical Magazine' for June 1832, Faraday published a translation of Nobili's first memoir, with notes, and later in the year he wrote a long letter to M. Gay-Lussac, on Nobili and Antinori's errors in magneto-electric induction. In this letter he says, ' These philosophers unfortunately had no other knowledge of my researches than the short letter which I wrote to M. Hachette, and not being careful to refer to my papers (though it appears to me they should have done so, under the circumstances), they have mistaken altogether the sense of a phrase relating to the beautiful observations of M. Arago ; they have presumed that I had not previously done that which they thought they had done themselves ; and finally, they advance what appears to me to be erroneous ideas of magneto-electric currents, and give their ideas as corrections of mine, which had not as yet come under their eyes.

'First, let me rectify that which I consider as the most serious error, the misinterpretation given to my words ; for those committed in the experiments would have been easily removed in the course of time.

'M. Nobili says : "He (Faraday) then (ten years ago) recognised, as the notice says, that by the rotation of a metallic disc, under the influence of a magnet, we may produce electric currents in the direction of the radii of the disc, in sufficient quantity to make this disc become a new electric machine." Now I never said that which is here imputed to me. I said " the extraordinary effect discovered by M. Arago was connected in its nature with the electro-magnetic rotation, which I had discovered several years before." I never said, and never had the intention of saying, that I " had discovered that which M. Arago discovered." I have the most earnest desire to have this error removed, for I have always admired the prudence and philosophic reserve shown by M. Arago, in resisting the temptation to give a theory of the effect he had discovered, so long as he could not devise one which was perfect in its application, and in refusing his assent to the imperfect theories of others.'

Faraday then continues : ' These philosophers say, "We have recently verified, extended, and perhaps rectified in some parts, the results of the English philosopher." With the greatest desire to be corrected when in error, it is still impossible for me to discover in the writings of these gentlemen any correction by which I can profit.' And then at great length he examines and compares their results with his own, and concludes thus :—

' I cannot terminate this letter without again expressing the regret I feel in having been obliged to write it. But if it be remembered that the memoirs of the Italian philosophers were written and published *after* my original papers, that their last writing has appeared in the same number of the " Annales de Chimie et de Physique " with mine ; and that, consequently, they have the *appearance* of carrying science beyond that which I had myself done ; that both their papers accuse me of errors in experiment and theory, and, beyond that, of good faith ; that the last of these writings bears the date of March, and has not been followed by any correction or retractation on the part of the authors, though we are now in December ; and that I sent them, several months ago, copies of my original papers, and also copies of notes on a translation of their first paper : and if it be remembered that,

after all, I have none of those errors to answer for
with which they reproach me, and that the memoirs of
these gentlemen are so worded, that I was constrained
to reply to the objections they made against me; I
hope that no person will say that I have been too
hasty to write that which might have been avoided;
or that I should have shown my respect for the truth,
or rendered justice to my own writings, and this branch
of science, if, knowing of such important errors, I had
not pointed them out.

'I am, my dear Sir, yours very faithfully,

'M. FARADAY.'

The records of 1832–33–34 that now remain, show
the vast amount, and the high importance of the
work which Faraday did. But they show very little
of the reputation which he gained, and still less of his
nature.

I.

It will be well to divide his work into that which he
did for the Royal Society; that which he did for the
Royal Institution; and that which he published elsewhere.

The third series of 'Experimental Researches in Electricity' was on the identity of electricities derived from
different sources, and on the relation by measure of
common and voltaic electricity.

August 25th, 1832, he began his experiments. He
soon proved that ordinary (frictional) electricity affects
the galvanometer.

August 30th, 31st, September 1st and 3rd, he
worked on the chemical decompositions produced by

the common frictional electrical current. On the latter day he writes in his note-book, 'As identity of common and voltaic electricity is proved, we may reason from the former, when intense, as to the manner of action of the latter.'

He then goes to experiments on voltaic decomposition, which ultimately formed part of the fifth series of researches; and as early as September 8 he made an experiment on chemical decomposition without any poles.

September 14th, he experiments on the effect of tension: 'The number of Leyden jars (8 and 15) charged, measured the tension and the number of turns of the plate machine, the quantity of the electricity.'

He then made a standard voltaic arrangement of platina and zinc wire $\frac{1}{18}$th of an inch in diameter, and a standard acid of one drop of sulphuric acid in four ounces of water; and then he compares the voltaic action with the action of the plate machine on the galvanometer. And September 15 he works on chemical decomposition, and ends thus: 'Hence it would appear that both in magnetic deflection and in chemical effect the current of the standard voltaic battery for eight beats of the watch was equal to the electricity of thirty turns of the machine, and that therefore common and voltaic electricity are alike in all respects.'

The paper in which his own facts and all he could collect elsewhere on the subject are contained, was sent to the Royal Society December 15, and was read on the 10th and 17th of January. At the conclusion he says, 'The extension which the present investigations have enabled me to make, of the facts and views constituting the theory of electro-chemical decomposition, will,

with some other points of electrical doctrine, be almost
immediately submitted to the Royal Society in another
series of these researches.

The excellent summary which Dr. Tyndall has made
of the relation by measure of common and voltaic electricity must be mentioned here :—

'After he had proved to his own satisfaction the
identity of electricities, he tried to compare them
quantitatively together. The terms quantity and intensity, which Faraday constantly used, need a word of
explanation here. He might charge a single Leyden
jar by twenty turns of his machine, or he might charge
a battery of ten jars by the same number of turns.
The *quantity* in both cases would be sensibly the same,
but the *intensity* of the single jar would be the greatest,
for here the electricity would be less diffused. Faraday
first satisfied himself that the needle of his galvanometer was caused to swing through the same arc by
the same quantity of machine electricity, whether it
was condensed in a small battery or diffused over a
large one. Thus the electricity developed by thirty
turns of his machine produced, under very variable conditions of battery surface, the same deflection. Hence
he inferred the possibility of comparing, as regards
quantity, electricities which differ greatly from each
other in intensity.

'His object now is to compare frictional with voltaic
electricity. Moistening bibulous paper with the iodide
of potassium—a favourite test of his—and subjecting
it to the action of machine electricity, he decomposed
the iodide, and formed a brown spot where the iodine
was liberated. Then he immersed two wires, one of
zinc, the other of platinum, each $\frac{1}{13}$th of an inch in

diameter, to a depth of $\frac{5}{8}$ths of an inch in acidulated water during eight beats of his watch, or $\frac{3}{20}$ths of a second ; and found that the needle of his galvanometer swung through the same arc, and coloured his moistened paper to the same extent, as thirty turns of his large electrical machine. Twenty-eight turns of the machine produced an effect distinctly less than that produced by his two wires. Now, the quantity of water decomposed by the wires in this experiment totally eluded observation ; it was immeasurably small ; and still that amount of decomposition involved the development of a quantity of electric force which, if applied in a proper form, would kill a rat, and no man would like to bear it.

' In his subsequent researches " on the absolute quantity of electricity associated with the particles or atoms of matter," he endeavours to give an idea of the amount of electrical force involved in the decomposition of a single grain of water. He is almost afraid to mention it, for he estimates it at 800,000 discharges of his large Leyden battery. This, if concentrated in a single discharge, would be equal to a very great flash of lightning ; while the chemical action of a single grain of water on four grains of zinc would yield electricity equal in quantity to a powerful thunderstorm. Thus his mind rises from the minute to the vast, expanding involuntarily from the smallest laboratory fact till it embraces the largest and grandest natural phenomena.'

The fourth series of researches was on a new law of electric conduction and on conducting power generally. It was received at the Royal Society April 24.

December 24th, 1832, Faraday says in his note-book, 'Can an electric current, voltaic or not, decompose a solid body—ice, &c.? If it cannot, what would frozen gum, lac, wax, &c.?'

January 23, 1833, he begins his experiments on ice. The ice was not quite dry, and so the needle was deflected. On the 24th he says, 'Made some excellent experiments on ice—quite dry; at 10°, or perhaps under; not the slightest deflection of the needle occurred.' On the 26th, 'If ice will not conduct, is it because it cannot decompose?'

His paper begins thus:—'I was working with ice, when I was suddenly stopped by finding that ice was a non-conductor of electricity.'

In a manuscript note to this paper, Faraday says:— '"Franklin's Experiments on Electricity," 4to, 5th edition, 1774, p. 36: "A dry cake of ice or an icicle held between two (persons) in a circle likewise prevents the shock, which one would not expect, as water conducts it so perfectly well."'

February 14th he began to experiment 'on substances solid at common temperatures, but fusible, and of such composition as was presumed would supply the place of or act like water.'

Next he took nitre: 'Whilst nitre was solid it did not conduct, i.e. no current passed through it affecting the galvanometer; on melting the nitre, and then putting the negative pole on the galvanometer, the needle was *knocked round*, as if the metals had touched through the nitre, and strong decomposing action took place. On allowing temperature to fall, the moment nitre solidified, the current through it ceased, yet negative wire was actually imbedded and cemented in the nitre.'

'Hence,' he says, 'nitre is exactly like water: whilst

solid it is a non-conductor, and when fluid a *conductor
and decomposed*; and therefore water has no peculiar
distinction, or action, or any exclusive power, in voltaic
chemical decomposition.

February 15th he thus enters his new law :—'This
general assumption of insulating powers, so soon as
fluid matter becomes solid, a new point, before un-
suspected, and very extraordinary. Seems to confer a
new property on the matter in the second state. Curious
that as gas and as solid non-conduct, and that as liquid
conduct.'

'This assumption of two states, perhaps, connected
with the conducting power of carbon, and non-conducting power of diamond.'

'Does not insulation by solid show that decomposition by voltaic pile is due to slight power superadded
upon previous chemical attractive forces of particles
when fluid?   Since mere fixation of particles prevents,
it must be slight.'

'Does it not show very important relations between
the decomposibility of such bodies and their conducting
power?   As if here the electricity were only a transfer
of a series of alternations or vibrations, and not a body
transmitted directly.   May settle or relate to question
of materiality or fluid of electricity.'

On February 21 he experimented first with sulphuret
of silver.   '*Very extraordinary*.'

'When all was cold, conducted a little (by the galvanometer).   *The heat rose as the conducting power increased* (a curious fact); yet I do not think it became
high enough to *fuse the sulphuret*.   The whole passed
whilst in the solid state.   The hot sulphuret seems to
conduct as a metal would, and could get sparks with

wires at the end, and a fine spark with charcoal.' And then he proceeds to examine a multitude of other substances.

February 26th he writes, ' Chloride of magnesium, when solid, and wire freezed in, non-conductor. When fused, conducted very well, and was decomposed. At P. pole much action and gas, Chlorine (?). At N. pole magnesium separated and no gas. Sometimes magnesium burnt, flying off in globules, burning brilliantly. When wire at that pole put in water or dilute muriatic acid, matter round it acted powerfully, evolving hydrogen and forming magnesia; and when wire and surrounding matter held in spirit lamp, *magnesium* burnt with intense light into *magnesia*. VERY GOOD EXPERIMENT.'

April 1st, he returns to the sulphuret of silver again. ' All the effects of electro-chemical decomposition seem to show that in ordinary chemical affinity the particles exert an influence not merely on those with which they are combined, but also, although to a weaker extent, upon those particles combined with their neighbours: that, in fact, it is not a mere tendency to unite particle to particle, but that tendency is general, and that even those in excess exert an influence, though it be not enough to overpower definite combination. Many facts in chemistry also bear on this view, that particles act in common. Berthollet, Phillips, &c., have quoted cases, but it is not merely incidental in these phenomena. Electro-chemical decomposition seems to be *essentially dependent* upon it.'

April 5th, he still worked on the sulphuret of silver, and says : ' Hence it is quite clear that a solid can conduct, that it can decompose whilst solid, that

increasing heat *increases* conducting power, that ele-   1833.
ments are electro-chemically arranged, that sulphur is   Æt. 41.
either positive or negative (as folks say) at pleasure.'
April 13th.—'Why did Davy require water in decomposing potassa?'

'If decomposition by voltaic battery depended upon
the attraction of the poles being stronger than that of
the particles separated, it would follow that the *weakest
electrical* attraction was stronger than the strongest, or
than very strong, *chemical* attraction; i.e. such as exists
between oxygen and hydrogen, acid and alkali, potassium and oxygen, chlorine and sodium, &c. This not
likely.'

'If voltaic decomposition of the kind I believe, then
revise all substances upon the new view, to see if they
may not be decomposed, &c.'

'A single element is never attracted by a pole,
i.e. without attraction of other element at other pole.
Hence doubt Mr. Brande's experiments on attraction
of gases and vapours. Doubt attraction by poles altogether.'

Professor Tyndall sums up this fourth series of experimental researches in electricity thus:—'He found that
though the current passed through water, it did not pass
through ice; why not, since they are one and the same
substance? Some years subsequently he answered this
question by saying that the liquid condition enables
the molecule of water to turn round so as to place itself
in the proper line of polarisation, while the rigidity of
the solid condition prevents this arrangement. This
polar arrangement must precede decomposition, and
decomposition is an accompaniment of conduction. He
then passed on to other substances; to oxides and

chlorides, and iodides, and salts, and sulphurets, and found them all insulators when solid, and conductors when fused. In all cases, moreover, except one—and this exception he thought might be apparent only— he found the passage of the current across the fused compound to be accompanied by its decomposition. Is then the act of decomposition essential to the act of conduction in these bodies? Even recently this question was warmly contested. Faraday was very cautious latterly in expressing himself upon this subject; but as a matter of fact he held that an infinitesimal quantity of electricity might pass through a compound liquid without producing its decomposition. De la Rive, who has been a great worker on the chemical phenomena of the pile, is very emphatic on the other side. Experiment, according to him and others, establishes in the most conclusive manner that no trace of electricity can pass through a liquid compound without producing its equivalent decomposition.'

The fifth series was on electro-chemical decomposition; new conditions of electro-chemical decomposition, influence of water in electro-chemical decomposition, and theory of electro-chemical decomposition. It was received at the Royal Society June 18. This series is continued in the seventh series on electro-chemical decomposition (continued), and is so connected with the eighth series on the electricity of the voltaic pile, that these three papers must be considered as one vast work. In the two first, Faraday tries to make clear to himself what actually takes place in solutions through which currents of electricity are passing, and in the third paper he applies the facts he had obtained, and proves that they hold good in the voltaic pile.

Having satisfied himself of the identity of the different electricities, and of their difference only in intensity, he thought it probable that the most intense would, when applied to chemical decomposition, give new facts and new views. In April he passes the machine electricity through pieces of litmus and turmeric moistened and connected by solution of sulphate of soda. Wherever the current entered or left the test paper, there was evidence of decomposition, 'indicating at once an internal action of the parts suffering decomposition, and appearing to show that the power that is effectual in separating the elements is exerted there and not in the poles.'

May 2, 1833, his note-book shows an inquiry of the greatest interest as regards his researches on light in 1845. He begins, 'As to effect of decomposing solution on polarised ray of light. It can be only two directions, one across the current, the other along it.' 'Have been passing ray of polarised light through decomposing solutions to ascertain if any sensible effect on the light.'

Saturated solution of sulphate of soda was first used, and the polarised ray passed through an extent of seven inches *across*, and afterwards *in*, the direction of the electric current.

'On making or breaking contact not the slightest effect could be perceived on the polarised ray.'

'I do not think, therefore, that decomposing solutions or substances will be found to have (as a consequence of decomposition or arrangement for the time) any effect on the polarised ray.'

'Should now try non-decomposing bodies, as solid nitre, nitrate of silver, borax, glass, &c., whilst solid, to see if any internal state induced, which by decomposition is destroyed, i.e. whether, when they cannot decompose, any state of electrical tension is present. My borate of lead glass good, and common electricity better than voltaic.'

May 6 he makes further experiments, and concludes, ' Hence I see no reason to expect that any kind of structure or tension can be rendered evident, either in decomposing or non-decomposing bodies, in insulating or conducting states.'

He then goes on with his experiments on decomposition.

And May 16 he writes, ' Is the law this? " Equal currents of electricity measured by the galvanometer evolve equal volumes of gas or effect equal chemical actions in a constant medium ? " Is it possible it may generalise so far as to give equal chemical action, estimated on the same elements on variable media ? Ought it not to be so if decomposition essential to conduction ?' And then he proceeds to experiment with different sized poles, different decomposing solutions, and different kinds of poles, including water as a pole.

In his paper he sums up his conclusion as to the nature of electro-chemical decomposition thus : ' It appears to me that the effect is produced by an internal corpuscular action, exerted according to the direction of the electric current, and that it is due to a force either *superadded to* or *giving direction to the ordinary chemical affinity* of the bodies present. The body under decomposition may be considered as a mass of acting particles, all those which are included in the course of the electric current contributing to the final effect.'

'The poles are merely the surfaces or doors by which the electricity enters into or passes out of the substances suffering decomposition. They limit the extent of that substance in the course of the electric current, being its *terminations* in that direction. Hence the elements evolved pass so far and no further.'

Dr. Tyndall's account of the continuation of the fifth series gives a far clearer view than can be gathered from the notes of the experiments. This research lasted all the autumn of 1833.

Dr. Tyndall says, 'His paper on electro-chemical decomposition, received by the Royal Society January 9, 1834, opens with the proposal of a new terminology. He would avoid the word "current" if he could. He does abandon the word "poles" as applied to the ends of a decomposing cell, because it suggests the idea of attraction, substituting for it the perfectly neutral term *electrodes*. He applied the term *electrolyte* to every substance which can be decomposed by the current, and the act of decomposition he calls *electrolysis*. All these terms have become current in science. He called the positive electrode the *anode*, and the negative one the *cathode*, but these terms, though frequently used, have not enjoyed the same currency as the others. The terms *anion* and *cation*, which he applied to the constituents of the decomposed electrolyte, and the term *ion*, which included both anions and cations, are still less frequently employed.

'Faraday now passes from terminology to research; he sees the necessity of quantitative determinations, and seeks to supply himself with a measure of voltaic electricity. This he finds in the quantity of water decomposed by the current. He tests this measure in all possible ways, to assure himself that no error can arise from its employment. He places in the course of one and the same current a series of cells with electrodes of different sizes, some of them plates of platinum, others merely platinum wires, and collects the gas liberated on each distinct pair of electrodes. He finds the quantity of gas to be the same for all. Thus he concludes that when the same quantity of electricity is caused to pass through a series of cells containing acidulated water, the electro-chemical action is independent of the size of the electrodes. He next proves that variations in intensity do not interfere with this equality of action. Whether his battery is charged with strong acid or with weak; whether it consists of five pairs or of fifty pairs; in short, whatever be its source, when the same current is sent through his series of cells, the same amount of decomposition takes place in all. He next assures himself that the strength or weakness of his dilute acid does not interfere with this law. Sending the same current through a series of cells containing mixtures of sulphuric acid and water of different strengths, he finds, however the proportion of acid to water might vary, the same amount of gas to be collected in all the cells. A crowd of facts of this character forced upon Faraday's mind the conclusion that the amount of electro-chemical decomposition depends, not upon the size of the electrodes, not upon the intensity of the current, not upon the strength of the solution, but solely upon the quantity of electricity which passes through the cell. The quantity of electricity he concludes is proportional to the amount of chemical action. On this law Faraday based the construction of his celebrated voltameter or measurer of voltaic electricity.

' But before he can apply this measure he must clear his ground of numerous possible sources of error. The decomposition of his acidulated water is certainly a *direct* result of the current ; but as the varied and important researches of MM. Becquerel, De la Rive, and others had shown, there are also *secondary* actions, which may materially interfere with and complicate the pure action of the current. These actions may occur in two ways : either the liberated *ion* may seize upon the electrode against which it is set free, forming a chemical compound with that electrode ; or it may seize upon the substance of the electrolyte itself, and thus introduce into the circuit chemical actions over and above those due to the current. Faraday subjected these secondary actions to an exhaustive examination. Instructed by his experiments, and rendered competent by them to distinguish between primary and secondary results, he proceeds to establish the doctrine of " definite electro-chemical decomposition."

' Into the same circuit he introduced his voltameter, which consisted of a graduated tube filled with acidulated water and provided with platinum plates for the decomposition of the water, and also a cell containing chloride of tin. Experiments already referred to had taught him that this substance, though an insulator when solid, is a conductor when fused, the passage of the current being always accompanied by the decomposition of the chloride. He wished now to ascertain what relation this decomposition bore to that of the water in his voltameter.

' Completing his circuit, he permitted the current to continue until " a reasonable quantity of gas " was collected in the voltameter. The circuit was then broken, and the quantity of tin liberated compared with the quantity of gas. The weight of the former was 3·2 grains, that of the latter 0·49742 of a grain. Oxygen, as you know, unites with hydrogen in the proportion of 8 to 1 to form water. Calling the equivalent, or, as it is sometimes called, the atomic weight of hydrogen 1, that of oxygen is 8; that of water is consequently 8 + 1, or 9. Now, if the quantity of water decomposed in Faraday's experiment be represented by the number 9, or in other words by the equivalent of water, then the quantity of tin liberated from the fused chloride is found by an easy calculation to be 57·9, which is almost exactly the chemical equivalent of tin. Thus both the water and the chloride were broken up in proportions expressed by their respective equivalents. The amount of electric force which wrenched asunder the constituents of the molecule of water was competent, and neither more nor less than competent, to wrench asunder the constituents of the molecules of the chloride of tin. The fact is typical. With the indications of his voltameter he compared the decomposition of other substances both singly and in series. He submitted his conclusions to numberless tests. He purposely introduced secondary actions. He endeavoured to hamper the fulfilment of those laws which it was the intense desire of his mind to see established. But from all these difficulties emerged the golden truth, that under every variety of circumstances the decompositions of the voltaic current are as definite in their character as those chemical combinations which gave birth to the atomic theory. This law of electro-chemical decomposition ranks, in point of importance, with that of definite combining proportions in chemistry.'

One note from his laboratory book and one from his paper may be added.

December 18th, 1833, he writes : 'The present voltaic apparatus, i.e. the trough, must be a very coarse, wasteful arrangement if referred to its first principle. For the zinc dissolved *ought* to supply electricity enough, if rightly collected, to affect the world almost.'

In his paper he says : 'Zinc and platina wires one-eighteenth of an inch in diameter and about half an inch long dipped into dilute sulphuric acid so weak that it is not sensibly sour to the tongue, or scarcely to our most delicate test-papers, will evolve more electricity in one-twentieth of a minute than any man would willingly allow to pass through his body at once. The chemical action of a grain of water upon four grains of zinc can evolve electricity equal in quantity to that of a powerful thunderstorm.'

December 29th, 1833, under the head of ' Electro-chemical equivalents—propositions relating to,' after considering the possibility of making a table of real electro-chemical equivalents, he continues, ' I must keep my researches really *experimental*, and not let them deserve anywhere the character of *hypothetical imaginations.*'

In the early part of 1834 Faraday was at work on the quantity of electricity evolved, and on secondary actions, and among other substances he used fluoride of lead. From this he worked for fluorine. On February 10th he writes : 'Daniell called on me to-day to ask me about my views of the elementary experiment of a single pair of metals, and the relation to poles, &c., &c., and if it had not occurred to me whether he might work at it. I told him my views, and wished him to work contemporaneously with me. He behaved very generously, leaving it open to me alone. But if another catches my idea, and works it out before I can write my paper, I shall always regret that Daniell has given way to me, and that another should come before him. Must leave fluorine, and hasten this matter of the VOLTAIC PILE. I showed Daniell my preparatory notes for the paper.' He then proceeds, on February 12th, to experiments on the generating plates and the intensity of the current they produce. First he uses amalgamated plates, and then he puts intervening platina plates ; and he writes, ' how very needful the current is *to decomposition* in the cases where the intervening platinas are used. But they cannot be cause and effect to each other. What is the common origin and cause of both? Must make this out. It is of no use continuing to suppose one as producing the other in either order.'

' These cases of retardation seem beautifully to show the antagonism of the chemical powers at the electro-motive parts with the chemical powers at the interposed parts. The first are producing electric effects, the second opposing electric effects, and the two seem equipoised as in a balance, and in both cause and effect appear to be identical with each other. Hence chemical action merely electrical action, and electric action merely chemical.'

Almost immediately he adds : ' I am continually wanting a clear, definite view of the actions in a single voltaic circuit.' Then again, after some further experiments on resistance, he says : 'Must consider the case of single decomposition very well and closely, for that includes the whole. Why is it necessary there should be a discharge of electricity before

1833-34.
Æt.41-43.

action can go on? Why not zinc alone decompose, and
how is it that in existing circumstances the platina
helps?'

On February 19th he makes the experiment, to show
that contact of the single pair of metals is not necessary
to produce chemical decomposition.

On the 22nd he writes : 'We seem to have the power
of deciding in certain cases of chemical affinity (as of
zinc with the oxygen of water, &c.), which of two modes
of action of the *one power* shall be exerted. In the
one mode we can transfer the power on it, being able
to produce elsewhere its equivalent of action; in
the other it is not transferred on, but exerted at the
spot. The first is the case of voltaic-electric production, the other the ordinary cases of chemical
affinity. But both are chemical actions, and due to one
power or principle. That no electricity is set free in
the latter case shows the equality of forces, and
therefore of electricity in those quantities which are
called chemical equivalents. Hence another proof
that chemical affinity and electricity are the same.'

He continues : 'I must very closely consider and
examine a case of combination in which no electric
current is produced, such as zinc in dilute sulphuric
acid, or oxide of lead in nitric acid, &c. What becomes
here of all the electricity which must pass during the
combination ? How is it destroyed between the particles? Of course they are able to neutralise each other,
but how do they neutralise ?

'Are not rubbed glass and the rubber exactly in the
state of zinc and the oxygen of water in an electro-
motive circle? i.e. when the rubbed glass and the
rubber are separated, are they not in the state assumed

1833-34. by the zinc and the oxygen before they combine,
Æt.41-43. and before the contact is made in a single voltaic
circle? They probably give an exalted view of the
conditions of the particles of the zinc and oxygen—
a permanent view, as it were. How do the states
agree?

'Would not this view, if supported, reduce both
modes of evolution to one common principle—the
mutual influence of neighbouring particles in the glass
not proceeding to a full effect; in the voltaic circle
being completed and being followed in succession by a
multitude of others of the same kind? In the last it is
the attraction of the zinc for the oxygen of the oxide,
and this would tell as well for the instances of induction, and perhaps of common electricity.'

He then experiments on the intensity of a current
and its power of affecting decompositions in different
resisting fluids; and March 8th writes : ' Hence I really
believe that the current passes, but the intensity is *not
sufficient* to cause decomposition of water.'

Faraday published these and other experiments in
the eighth series of his 'Researches.' It was received
at the Royal Society, April 7, 1834. It was on the
electricity of the voltaic pile : its source, quantity, intensity. In this and the two former papers Faraday
worked ' to remove doubtful knowledge ' regarding the
definite action of electricity on decomposing bodies, and
the identity of the power so used with the power to be
overcome. He got clear ideas of the absence of all
attractive power in the poles ; clear ideas of the active
state of each particle of the electroleids between the
eisode and exode ; clear ideas of the definite quantity of
chemical action caused by a definite quantity of electri-

city, and clear ideas that the contact of metals was not 1833–34.
Æt.41–43. the origin of the electro-motive force, but that volta-electric excitation and ordinary chemical affinity are ' both chemical actions and due to one force or principle.'

At the end of this paper he says : ' I would rather defer revising the whole theory of electro-chemical decomposition until I can obtain clearer views of the way in which the power under consideration can appear at one time as associated with particles giving them their chemical attraction, and at another as free electricity.'

The sixth series of ' Researches,' on the power of metals and other solids to induce the combination of gaseous bodies, was sent to the Royal Society, November 30, 1833. This paper arose from a fact observed in the course of one of the experiments, mentioned in the seventh series of ' Researches.' It furnishes the clearest picture of the way in which Faraday worked.

On September 17, 1833, he writes in his note-book : ' Have been comparing decomposition of muriatic acid

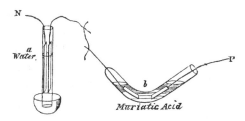

and water together, as to the equivalents of elements evolved by a given current of electricity.' ' I remarked that whilst b tube was being examined, the bulk of gas in *a* *had* diminished; therefore, put on *a* alone, and by battery evolved gas until it was full. Being left to itself, this gas gradually went or diminished, and after three or four hours, not a fourth part was left. At first, twelve o'clock, there were 116 parts, and at last, five o'clock, there were only 13·5 parts. Think this must have been an effect of permeability through the cork at top, by wires, &c., but must examine it closely, and also use tube hermetically sealed at the top.'

'September 18th.—To-day examined the 13·5 parts left yesterday; by heating spongy platina in it the gas diminished to two parts. Hence, think it cannot be due to permeability of cork, &c., for no sensible portion of air has entered. Think it must be due to recombination of the oxygen and hydrogen in some way.'

On 19th, after making more sure of his facts, he writes : 'I suspect all this is some combining power, possessed by the platina of the poles—perhaps given to it during the decomposition.' 'Must ascertain whether both poles, or only the positive has the power.' ' If poles have this power, the effect will immediately connect with that of spongy platina, and probably explain it.' 'Perhaps merely digesting platina in dilute sulphuric acid, or at least in nitro-muriatic acid, may give it this power.' 'Probably heating in air, or in flame with little muriate of ammonia vapour, or in chlorine, &c., will give this power to platina, in plate or lump. Probably also heat much assist it. Try all this.'

Almost immediately afterwards he writes : ' It is quite clear that the positive pole has peculiar power of

causing oxygen and hydrogen to combine.' And then
he left the subject till October 10, when he found that
positive pole, put into mixed gases, *became red-hot in
the middle part*, and the rest of the gas exploded.

' A pole, or, as it should now be called, a plate, was
merely heated by the spirit-lamp and blow-pipe, not
having been connected with the battery, and put up
into gas, oxygen and hydrogen. At first, there was no
action, but after a while, condensation began and went
on well at the last.'

October 11 he says : ' Hence heat can bring platina
into the acting state.' And then he tries mechanical
and chemical actions to prepare the platinum plate.

On October 14 he writes : ' Whilst heated and in
the sulphuric acid, the surfaces (of plates) acquired
such a state as to cause much friction when the pieces
were rubbed against each other. This no doubt
because of their perfectly clean state, and helps to
show that effect is due to that clean state of surfaces
which acid and battery induce.' Then he found that
by heating an active plate it sometimes lost its power,
and he writes : ' Must remember that platina can
combine with carbon by heat, and that probably the
surface is thus affected in these modes of igniting.'

This day he cleaned a platinum plate chemically
with potass, heat, sulphuric acid and water, and then
put it into the mixed gases. ' Instant excellent action ;
the gas rose quickly, the platinum became red-hot, and
Doberiner's effect was produced *without action of battery*
on the platina—*Good.*'

November 7, he tries the effect of gold and palladium, silver, copper. The two first acted.

November 8 and 12, tried the effect of mixing

other gases with the oxygen and hydrogen, and hydrogen alone.

Then he experimented on the rapidity with which
substances get dirty.

November 25, he says: 'Think this a subject of
great consequence, for I am convinced that the superficial actions of matter, and the actions of particles not
directly or strongly in combination, are becoming daily
more and more important in chemical as well as in
mechanical philosophy.'

The conclusion at which he arrived is thus stated in
his paper :—

'All the phenomena connected with this subject
press upon my mind the conviction, that the effects in
question are entirely incidental, and of a secondary
nature (not electrical, as M. Doberiner, the discoverer
of the action of spongy platinum, had considered),
that they are dependent upon the *natural conditions* of
gaseous elasticity, combined with the exertion of that
attractive force possessed by many bodies, especially
those which are solid, in an eminent degree, and probably belonging to all; by which they are drawn into
association more or less close, without at the same
time undergoing chemical combination, though often
assuming the condition of adhesion ; and which occasionally leads, under favourable circumstances, as in
the present instance, to the combination of bodies
simultaneously subjected to this attraction.'

In the abstract of the Friday evening discourse
which he gave on this subject in 1834, he writes :—

' The peculiar power of metallic bodies to produce such effects, he accounted for, first, by the supposition of their possessing a specific power of attraction for gases, totally different from chemical affinity ; and second, by the peculiar condition of elastic bodies when mixed. The first supposition he attempted to support thus : he threw a little magnesia on water, and at the same time filings of zinc on a different portion of water. The former immediately became wet and sank, the latter remained dry and floated ; in fact, it seemed to evince, as it were, a repulsive power towards the water. In the same manner, everyone knows that other metallic bodies are not easily wet. Immerse the blade of a knife in water ; on drawing it out it will not be equably wet, but the water will appear in patches, or run into globules. But suppose you dip in the platina plate, cleaned as directed, it comes out uniformly wet. Now the only difference is that the matters adhering to the surface have been in the latter case removed ; but they are chiefly gases, vapours, atmospheric air, &c. ; for such, therefore, metals must have a specific power of attraction, and, being thus in contact with them, refuse contact with liquid bodies. For the second point, Dalton has shown that the particles of one gas preserve under every pressure (short of that which produces liquefaction) the same relative distance ; but it appears that they may approach indefinitely near to the particles of any other gas. It is above shown that they may come into actual contact with a clean metallic surface. If then our platina plate be introduced into a mixture of oxygen and hydrogen gases, it is evident that an atom of each in an indefinitely near state of approximation, is at the same moment brought into contact with a solid substance; their elasticity is thus destroyed, but elasticity is the condition of their gaseous existence. This, therefore, can no longer continue; they combine, and fall down in the form of a liquid. Such,' said Mr. Faraday, 'is my theory; every one is of course partial to the child of his own imagination, and I have not, after much pains, been able to see where this is deficient. In submitting it to your attention, as the result of experiments conducted in your laboratory, I wish to show that I have not lightly prized or indolently reposed under the favours you have conferred on me by appointing me Fullerian Professor in this Institution. Should my views appear correct and satisfactory, they will receive their highest reward in your approbation; should they appear to anyone to require further proof, I hope that I shall never shrink from their fair and candid discussion. We can all here have but one object—the elucidation and confirmation of truth.'

During the summer of 1834 Faraday was experimenting on a new construction of the voltaic battery. This was the result of his experiments on electrochemical decomposition. In the autumn he was busy comparing his battery with Wollaston's battery, when in October Mr. William Jenkin showed him that a magneto-electric shock could be obtained from a single pair of plates. He had found that if the wire which surrounds an electro-magnet be used to join the plates of a single cell, a shock is felt each time contact is broken, provided the ends of the wire are grasped one in each hand. It had long been known that a bright electric spark occurred under the same circumstances.

On October 15 Faraday began his experiments, 'as

Mr. Jenkin does not intend to work out the result any
further.'[1]

November 14, he writes : ' Now, then, begin to see light. The phenomenon of increased spark is merely a case of the induction of electric currents. If a current be established in a wire, and another wire forming a complete circuit be placed parallel to it, at the moment the current in the first is stopped, it induces a current in the same direction in the second, itself then showing but a feeble spark. But if the second be away, it induces a current in its own wire in the same direction, producing a strong spark. The strong spark in the current when alone is therefore the equivalent of the current it can produce in a neighbouring wire when in company.'

' These effects show that every part of an electric current is acting by induction on the neighbouring parts of the same current, even in the *same wire*, and the *same part* of the wire.'

' Further investigations ended in identifying these effects with the phenomena of induction, which I had been fortunate enough to develope in the first series of

---

[1] Elsewhere he writes, ' The number of suggestions, hints for discovery, and propositions of various kinds offered to me very freely, and with perfect goodwill and simplicity on the part of the proposers for my exclusive investigation and final honour, is remarkably great, and it is no less remarkable that but for one exception—that of Mr. Jenkin—they have all been worthless.

'It is quite natural that, when a man first catches sight of an analogy or relation, however imperfect it may be, he should suppose it a new and unconsidered subject, but it is very rare that the same thing has not passed before through the mind of a veteran, and been dismissed as useless. I have, I think, universally found that the man whose mind was by nature or self-education fitted to make good and worthy suggestions was also the man both able and willing to work them out.

' The volunteers are serious embarrassments generally to the experienced philosopher.'

1832-34.
Æt.40-43.

these "Experimental Researches." ' The results of this investigation into the extra current were sent on December 18 to the Royal Society as the ninth series of ' Researches,' on the influence by induction of an electric current on itself.

He ended this inquiry in December, and made some more experiments on his voltaic battery, which he finished in the second week of January 1835.

His work also for the Royal Institution during 1832, 1833, and 1834 was considerable.

In 1832, on the Saturdays in June, he gave a course of five lectures, on some points of domestic chemical philosophy—a candle, a lamp, a chimney, a kettle, ashes. As an example of the mode in which he made his notes for lectures, those on the kettle are given in an appendix to this chapter. They well show his excessive neatness and exactness, and the superabundance of his illustrations. The notes for his experiments were put on one side of a sheet of paper, and the notes for his words on the opposite side; and for many years he kept to this arrangement.

His Friday discourses were: 1. On Dr. Johnson's Researches on the Reproductive Power of Planariæ (cut again and again, gain immortality under the knife). 2. Recent experimental Investigation of Volta-electric and Magneto-electric Induction. 3. Magneto-electric Induction, and the explanation it affords of Arago's Phenomena of Magnetism exhibited by moving Metals. 4. Evolution of Electricity, naturally and artificially, by the inductive action of the Earth's Magnetism. At the end of the notes of this lecture he says, ' Refer to Nobili's results and claim my own.' 5. On the Crispa-

tion of Fluids lying on vibrating Surfaces. And 6. On
Morden's Machinery for manufacturing Bramah's locks.
In 1833, in May and June, he gave a course of six
lectures on magnetism and electricity. His subjects
were Common Electricity, Voltaic Electricity, Thermo-
Electricity; Common Magnetism, Electro-Magnetism,
Magneto-Electricity —of this last lecture his notes begin,
'My own branch of science as to discovery, November
1831;' and he ends with general reflections on this
wonderful, universal, subtle, and Proteus-like power.

His Friday discourses were: 1. On the Identity of
Electricity derived from different sources. 2. On the
practical Prevention of Dry Rot in Timber. 3. On the
Investigation of the Velocity and Nature of the Electric
Spark and Light by Wheatstone. 4. On Mr. Brunel's
new mode of constructing Arches for Bridges. 5. On
the mutual Relations of Lime, Carbonic Acid, and
Water. 6. On a new Law of Electric Conduction.

In 1834, in May and June, he gave a course of six
lectures on the Mutual Relation of Electrical and
Chemical Phenomena; Chemical Action; Electrical Action; Association and mutual dependence of these two
modes of action in the voltaic pile; Electro-chemical decomposition; Combustion as an Electrical Phenomenon;
Relations of Chemical Affinity, Electricity, Heat, Magnetism, and other powers of Matter. In this last lecture
he gives his first utterance on the correlation of physical
forces: 'Now consider a little more generally the relation of all these powers. We cannot say that any one
is the cause of the others, but only that all are connected
and due to one common cause. As to the connection, observe the production of any one from another,
or the conversion of one into another.' Then he gives

experiments of conversion of chemical power into heat, of chemical power into electricity, of chemical power into magnetism; then of electrical power into heat, of electrical power into magnetism: of electrical power into chemical power; then of magnetism into heat, of magnetism into chemical power, and of magnetism into electricity; then of heat into magnetism. His notes continue thus: 'This relation is probably still more extended and inclusive of aggregation, for as elements change in these relations they change in those. Experiments: sub-carb. of potass and muriate of lime, solid; nitrate of ammonia and Glauber-salt, liquid. And even gravitation may perhaps be included. For as the local attraction of chemical affinity becomes attraction at a distance in the form of electricity and magnetism, so gravitation itself may be only another form of the same power.' In 1853, Faraday marked these notes and experiments with his initials, and added: 'Correlation of physical forces.' Mr. Grove's lecture at the London Institution was in 1842; Faraday's at the Royal Institution, June 21, 1834.

He gave four Friday discourses, the first on the principle and action of Ericsson's caloric engine. At the end of his notes of this lecture, he says: 'Must always work practically; never give a final opinion except on that.' The other lectures were on Electrochemical Decomposition; on the definite action of Electricity; and on new applications of the products of Caoutchouc.

In addition to his Royal Society Papers and his Institution Lectures, he published a paper in the 'Edinburgh New Philosophical Journal,' in 1833, on the Planariæ, and, in the same year, another in the 'Philo-

sophical Magazine,' on a means of preparing the 1832–34.
Organs of Respiration so as considerably to extend ÆT.40–43.
the time of holding the breath, with remarks on its
application in cases in which it is required to enter
an irrespirable atmosphere, and on the precautions
necessary to be observed in such cases.

In 1832 he published, in the 'Philosophical Magazine,' notes on Signor Nobili's paper, and also on Signor
Negri's magneto-electric experiments; and in 1834 a
paper on the magneto-electric spark and shock, and on
a peculiar condition of electric and magneto-electric
conduction.

II.

His reputation shows itself in the titles and appoint-
ments which he received. In 1832 he was made Hon.
Member of Philadelphia College of Pharmacy; and of
the Chemical and Physical Society, Paris; Fellow of
the American Academy of Arts and Sciences, Boston;
Member of the Royal Society of Sciences, Copenhagen;
D.C.L. of Oxford University; and he received the
Copley medal from the Royal Society.

Regarding his degree at Oxford Dr. Daubeny wrote
to him:—

'My dear Sir,—There has been some talk amongst the
friends of science in this University about soliciting the
Heads of Colleges to propose Honorary Degrees for a
few of the most distinguished persons who are expected
here at the approaching meeting (of the British Association); but before a list is made out we are desirous of
ascertaining whether we are to expect the pleasure of
your attendance during any part of the time, and

whether such a mark of distinction would be acceptable
to you, as well as gratifying to the body who would
confer this mark of their consideration.

'I cannot help flattering myself that as when I last
saw you you told me you had not entirely given up
the intention of coming, the circumstance I have stated
may decide you in favour of being here, and that you
would value this tribute to your services to science the
more as coming from a body of men by whom such
honours to men of science have hitherto been but
rarely paid.

'Pray inform me in the course of the week whether
in the event of your honorary degree of D.C.L. being
determined on, we may reckon on your attendance for
a day or two days in the week of meeting.

'And believe me, yours very truly,
'CHARLES DAUBENY.'

In 1832, in December, the Royal Institution being
in trouble, a committee reported on all the salaries.
'The Committee are certainly of opinion that no reduc-
tion can be made in Mr. Faraday's salary, 100l. per
annum, house, coals, and candles; and beg to express
their regret that the circumstances of the Institution
are not such as to justify their proposing such an increase of it as the variety of duties which Mr. Faraday
has to perform, and the zeal and ability with which he
performs them, appear to merit.'

In 1833, in the earlier part of the year, Mr. Fuller
had founded a professorship of chemistry at the Royal
Institution, with a salary of about 100l. a year. Mr.
Faraday was appointed for his life, with the privilege

of giving no lectures. He was made Corresponding 1833. Member of the Royal Academy of Sciences of Berlin, Æт.41–42. and Hon. Member of the Hull Philosophical Society. In 1834 he was made Foreign Corresponding Member of the Academy of Sciences and Literature of Palermo.

The following letters from a French, a Dutch, and a German correspondent will show how his reputation was rising abroad.

M. HACHETTE TO FARADAY.

Paris: 30 août 1833.

'Monsieur et très-cher Confrère,—Ayant fait un petit voyage vers ma ville natale (Mézières, Ardennes), j'ai trouvé à mon retour votre lettre du 17 juin et la troisième série de vos recherches électriques, qui sera bientôt suivie d'une quatrième série. En lisant la dernière page (64) de votre mémoire sur la troisième série, j'ai remarqué ces mots : " *My first unfortunate letter to M. Hachette.*" Je vous avoue que je ne puis concevoir rien de malheureux dans l'annonce d'une découverte qui vous place au rang des plus heureux des grands physiciens.

'Supposez pour un moment qu'un membre de la Société Royale ait fait part de votre lecture dans une lettre particulière, que cette lettre soit tombée par hasard entre les mains d'un physicien qui aurait répété vos expériences et qui aurait la prétention de les avoir inventées. Vous pourriez alors dire, " *Oh, unfortunate letter!*" Tout le contraire arrive ; vous annoncez une grande découverte à la Société Royale ; elle est transmise de suite à Paris en votre nom ; toute la gloire de

E 2

1833.

ÆT.41–42.

l'inventeur vous est assurée. Que pouvez-vous désirer
de plus ?—que personne ne travaille dans la mine que
vous aviez ouverte ? cela est impossible. L'excellence
de votre découverte est en raison des efforts que chaque
physicien fera pour l'étendre. Ne lisez-vous pas avec
plaisir la feuille du journal français, " Le Temps," qui
chaque mercredi donne la séance de l'Académie des
Sciences du lundi précédent ? Y a-t-il un académicien
français qui se plaigne de cette prompte communication ? Non ; c'est le contraire. Les routes en fer ne
donnent pas encore aux nouvelles scientifiques la vitesse
de la pensée, et sous ce rapport elles sont encore très-
imparfaites.

'Je ne me pardonnerais pas si j'avais à me reprocher
une communication qui aurait été pour vous la cause
d'un vrai malheur ; mais permettez-moi de vous dire
qu'en cette circonstance le malheur n'est que dans votre
imagination. Je crois avoir contribué à étendre votre
gloire, votre renommée, à vous rendre la justice qui
vous est due, et je m'en félicite. J'espère que vous
partagerez mes sentiments.

'Agréez l'assurance de mon bien-sincère attachement.

'HACHETTE.'

Professor Mohl gives an account of the first electric
telegraph in Germany, in the letter in which he states
his opinion of the value of Faraday's researches.

PROFESSOR MOHL TO FARADAY.

'Utrecht : November 15, 1833.

'I made an excursion to Germany, and visited the
celebrated University of Göttingen.

'If I am not mistaken, the lectures of old Blumenbach would scarcely draw an audience to the lecture-room of a mechanics' institution. Their library, however, it must be said, is excellent, and challenges, as far as usefulness is concerned, any other in existence. Gauss has got up a very neat apparatus, a sort of magnetic telegraph. Two bar-magnets of a pound weight are suspended in different places at a distance of about one and a half mile (English), each has wires coiled round but not touching them; these wires communicate through the open air over roofs and steeples, and the action of a couple of galvanic plates in one place gives motion to the magnet placed at the distance of one and a half mile. The thing, at any rate, is very curious.

'I have read with great pleasure your new series of experiments. They are sure to carry your name down to posterity, as long as there will be anything existing like science.'

BARON HUMBOLDT TO FARADAY.

'Berlin: 28 mai 1834.

'Monsieur,—Une prédilection qui date de bien loin me ramène sans cesse aux admirables découvertes dont vous avez enrichi les sciences physiques. Votre mémoire sur l'identité des prétendus différents genres d'électricité ("Phil. Trans." for 1833, Part I. p. 47), mémoire remarquable par l'esprit philosophique qui l'a dicté, m'a encore occupé ces derniers jours, et j'y trouve quelques doutes sur les apparences lumineuses que Walsh et Ingenhauss assurent avoir observées. Vous avez fait d'inutiles recherches pour trouver où ces observations et celles de Fahlberg, qui a vu des étincelles semblables à la décharge d'une bouteille de Leyde (une vive lumière) se trouvent consignées? Je suis heureux de pouvoir les indiquer..... Je répète que dans les expériences que j'ai faites sur les gymnotes en Amérique et à Paris, où l'animal n'a vécu que quelques jours, je n'ai pu voir ces étincelles dont Ingenhauss a été le témoin. Je n'ai pas non plus vu le poisson lancer le coup de loin à travers des couches d'eau pour tuer des poissons qu'il voulait avaler, ce qui indiquerait une tension électrique d'une force extrême. Ce dernier fait a été vu souvent en Suède par Fahlberg. Malgré le beau travail de M. le docteur Davy sur les torpilles, il serait bien à désirer que votre Société Royale fît venir à Londres des gymnotes vivants (six ou huit au moins), assez faciles à transporter dans la courte navigation de la Guyane en Angleterre. Je suis persuadé qu'avec les connaissances électro-magnétiques et physiologiques que nous possédons aujourd'hui, l'étude des phénomènes du gymnote devrait répandre une vive lumière sur les fonctions des nerfs et de mouvement musculaire dans l'homme.

'J'ai déjà observé qu'en coupant un gymnote en deux portions, il n'y a que celle où je trouve le cerveau et le cœur qui continue à donner des coups électriques ("Voyage," t. ii. p. 182). L'organe électrique n'agit que sous l'influence immédiate du cerveau (et du cœur?). Un des plus grands mystères physiologiques me paraît être la ligature d'un nerf. Je suis de nouveau occupé d'expériences galvaniques sur les effets de la ligature lorsque le nerf (c'est-à-dire le faisceau de fibres médullaires que nous appelons nerf) a été lacéré. La ligature n'a jamais été essayée dans un gymnote—je veux dire la ligature de l'organe électrique.

'Je vous supplie, monsieur, de pardonner l'ennui de ces lignes, et d'agréer l'hommage de la haute et respec-

tueuse considération qui est due à vos travaux et à 1834.
votre noble dévouement pour les sciences. Æт. 42.
' Votre très-humble et très-obéissant serviteur,
' LE BARON DE HUMBOLDT.'

III.

But few marks of his nature during 1832, 1833, and 1834 remain.

In 1832 he collected and bound together the different papers, notes, notices, &c., published in octavo up to this year, and he added this very characteristic preface :—

' Papers of mine published in octavo in the " Quarterly Journal of Science " and elsewhere, since the time that Sir H. Davy encouraged me to write the " Analysis of Caustic Lime." Some I think (at this date) are good, others moderate, and some bad. But I have put *all* into the volume, because of the utility they have been to me, and none more than the bad, in pointing out to me in future, or rather after times, the faults it became me to watch and avoid. As I never looked over one of my papers a year after it was written without believing, both in philosophy and manner, it would have been much better done, I still hope this collection may be of great use to me.'

The kind words with which he encouraged a youth of twenty-two years of age, who sent him some scientific work, is seen in a letter to Matteucci, who was then living in his native place, Forli.

FARADAY TO C. MATTEUCCI.

' Royal Institution : October 1, 1833.

' Sir,—I am very much your debtor for your kindness in sending me your papers and for your good opinion.

All such marks of goodwill are stimuli to me, urging
me still forward in the course which has obtained such
commendation.

.   .   .   .   .   .

' Being convinced you cannot refrain from pursuing
science by experiment, I need not express a hope that
you will do so manfully. No man of judgment can
work without succeeding, and you are not likely to
leave a course which has already made your name
known throughout the European Continent.

' Ever your obedient servant,

' M. FARADAY.'

The year 1835 is remarkable in Faraday's life, not so
much for the work he did, or for the marks of the
reputation he had gained, as for the character which
he showed in refusing and in accepting a pension from
the Prime Minister.

I.

On April 20 Sir James South wrote to him to say
that he would have a letter from Sir Robert Peel
acquainting him with the fact that, had Sir R. Peel
remained in office, a pension would have been given
him. On the 23rd he wrote to Sir James South, ' I
hope you will not think that I am unconscious of the
good you meant me, or undervalue your great exertions
for me, when I say that I cannot accept a pension
whilst I am able to work for my living. Do not from
this draw any sudden conclusion that my opinions are
such and such. I think that Government is right in
rewarding and sustaining science. I am willing to
think, since such approbation has been intended me,
that my humble exertions have been worthy, and I
think that scientific men are not wrong in accepting

the pensions ; but still I may not take a pay which is
not for services performed whilst I am able to live by
my labours.'

Changing his opinion in consequence of the judgment
of his father-in-law, for which he had the highest
regard and respect, Faraday sent another letter in the
place of this. It contained a less decisive refusal. He
heard no more of the pension until October 26, when
he was asked to wait upon Lord Melbourne, who was
then Prime Minister, at the Treasury. A conversation
took place, in which Lord Melbourne says he expressed
himself ' certainly in an imperfect and perhaps in too
blunt and inconsiderate a manner.' It is probable that
he also said that he looked upon the whole system of
giving pensions to literary and scientific persons as a
piece of humbug.

The same evening (that is, the day of the conversation) Faraday left this note, with his card, at Lord
Melbourne's office.

'TO THE RIGHT HON. LORD VISCOUNT MELBOURNE, FIRST
LORD OF THE TREASURY.

'October 26.

' My Lord,—The conversation with which your Lordship honoured me this afternoon, including, as it did,
your Lordship's opinion of the general character of the
pensions given of late to scientific persons, induces me
respectfully to decline the favour which 1 believe your
Lordship intends for me ; for I feel that I could not,
with satisfaction to myself, accept at your Lordship's
hands that which, though it has the form of approbation, is of the character which your Lordship so pithily
applied to it.'

Sir James South lived near Holland House, and

occasionally he saw Miss Fox and others in his observatory, and from thence probably the suggestion to Lord Melbourne of a pension to Faraday originally proceeded. When Faraday refused the offer, he explained why he did so to Sir James South, and through him to Miss Fox. This he did on November 6, in two letters which show that no bad temper was in him. To Sir James South he says, 'I hope that in doing what was right I have not given others occasion to have one evil thought of me. Since I first knew of the affair, nothing has been nearer to my mind than the desire, whilst I preserved my self-respect, to give no one occasion of offence.' To Miss Fox he says, ' I shall never forget that what you know of me thus far has gained your approbation, and it will be doubly my desire henceforward to deserve and retain it.'

The refusal of the pension became known, and it even reached the King, and it pleased him to remind his Prime Minister of it whenever he had an opportunity. Perhaps to avoid these remarks, and perhaps for other reasons, 'an excellent lady, who was a friend both to Faraday and the Minister, tried to arrange matters between them ; but she found Faraday very difficult to move from the position he had assumed. After many fruitless efforts, she at length begged of him to state what he would require of Lord Melbourne to induce him to change his mind. He replied, ' I should require from his Lordship what I have no right or reason to expect that he would grant—a written apology for the words he permitted himself to use to me.' [1]

---

[1] Dr. Tyndall uses these words from his recollection of a conversation with Faraday.

A letter in which he ' returns his heartfelt thanks
to Lady Mary Fox for all the kindness she had shown
him,' is dated the day previous to that on which Lord
Melbourne wrote the following letter :—

LORD MELBOURNE TO FARADAY.

' November 24.

' Sir,—It was with much concern that I received
your letter declining the offer which I considered
myself to have made in the interview which I had with
you in Downing Street ; and it was with still greater pain
that I collected from that letter that your determination
was founded upon the certainly imperfect, and perhaps
too blunt and inconsiderate manner in which I had
expressed myself in our conversation. I am not unwilling to admit that anything in the nature of censure
upon any party ought to have been abstained from
upon such an occasion ; but I can assure you that my
observations were intended only to guard myself against
the imputation of having any political advantage in
view, and not in any respect to apply to the conduct of
those who had or hereafter might avail themselves of a
similar offer. I intended to convey that, although I
did not entirely approve of the motives which appeared
to me to have dictated some recent grants, yet that
your scientific character was so eminent and unques-
tionable as entirely to do away with any objection which
I might otherwise have felt, and to render it impossible
that a distinction so bestowed could be ascribed to any
other motive than a desire to reward acknowledged
desert and to advance the interest of philosophy.

' I cannot help entertaining a hope that this explanation may be sufficient to remove any unpleasant or

unfavourable impression which may have been left upon your mind, and that I shall have the satisfaction of receiving your consent to my advising His Majesty to grant to you a pension equal in amount to that which has been conferred upon Professor Airy and other persons of distinction in science and literature.'

The same day Faraday wrote :—'My Lord, your Lordship's letter, which I have just had the honour to receive, has occasioned me both pain and pleasure— pain, because I should have been the cause of your Lordship's writing such an one, and pleasure, because it assures me that I am not unworthy of your Lord- ship's regard.

' As, then, your Lordship feels that, by conferring on me the mark of approbation hinted at in your letter, you will be at once discharging your duty as First Minister of the Crown, and performing an act conso- nant with your own kind feelings, I hesitate not to say I shall receive your Lordship's offer both with pleasure and with pride.'

This gentle letter should have brought this affair of the pension to an end, but unfortunately four days afterwards an account full of error respecting the con- versation between Lord Melbourne and Faraday was published in the 'Times' of Saturday, November 28, under the head of 'Tory and Whig patronage to Science and Literature.'

It was copied from 'Fraser's Magazine :'—

'*Mr. F.* I am here, my Lord, by your desire; am I to understand that it is on the business which I have partially discussed with Mr. Young? (Lord M.'s Secre- tary.) *Lord Melbourne.* You mean the pension, don't you? *Mr. F.* Yes, my Lord. *Lord M.* Yes, you mean the pension, and I mean the pension too. I hate the name of the pension. I look upon the whole system of giving pensions to literary and scientific persons as a piece of gross humbug. It was not done for any good purpose, and never ought to have been done. It is a gross humbug from beginning to end. *Mr. F. (rising and making a bow).* After all this, my Lord, I perceive that my business with your Lordship is ended. I wish you a good morning.'

The day after the article in the ' Times ' appeared, November 29, Dr. Holland wrote to Faraday at the request of Lord Melbourne, to say that the King expressed great satisfaction in the arrangement made as to the pension. ' We spoke of the paper which has appeared in the " Times ;" he begs you not to be disquieted by this in reference to him. Though regretting the circumstance, he was prepared for the likelihood of these things being converted to party purposes.' He thought it best that no public notice should be taken by Faraday of the paper.

To Lord Holland, however, it appeared otherwise ; and Faraday, hearing of this from Dr. Holland, wrote to Lord Holland to say, that if Lord Melbourne wishes it ' Sir James South with Dr. Holland will take such steps as under existing circumstances appear to be most expedient. Urged by Lord Holland, Dr. Holland then suggested that Faraday should write a very brief letter. Faraday replies, that he cannot proceed without Sir James South, and adds, ' The pension is a matter of indifference to me, but other results, some of which have already come to pass, are not so. The continued

renewal of this affair, to my mind, tempts me at times
to what might be thought very ungenerous under the
circumstances—namely, even at this late hour, a determined refusal of the whole.'

The next day Faraday had a conversation with a
gentleman regarding the article in the 'Times.' An account of this was published, with the signature ' T. R. S.,'
in the 'Chronicle' for December 8. The writer says, ' I
expressed to him (Faraday) my total disbelief in its
correctness. His answer was, " I am glad you don't
believe it. It is full of falsehood, and evidently written
for a factious purpose. You have my authority for
saying that." '

Six days after this talk, Faraday speaks of it thus, ' I
cannot now recollect whether I used the word falsehood. I had thought my expressions in reference to this
subject over beforehand, and had used the phrase " full
of error " to others.'

The 'Courier' of December 7, in a leading article
upon the imaginary conversation published in the
'Times,' said : 'The person most insulted and most
injured by it is Mr. Faraday, who will be suspected by
the whole kingdom, unless he contradict it in his own
name, of having authorised the publication and supplied
the scanty proportion of truth which was woven into
the web of fiction. It is not just to himself to allow a
whole month to elapse before he clears himself (in the
next number of " Fraser's Magazine ") of the suspicions
now entertained.'

Faraday the next day published a letter in the
'Times,' in which he says, ' I beg leave thus publicly
to state that neither directly nor indirectly did I communicate to the editor of " Fraser's Magazine " the

information on which that article (an extract of which was published in the "Times" of the 28th) was founded, or further, either directly or indirectly, any information to or for any publication whatsoever.'

On December 12, Miss Fox urges him to make a public vindication of Lord Melbourne. To this he answers the same day, 'I am persuaded that any statement of mine in the public papers cannot NOW be attended with any advantage to either party.' (The draft of this letter is in Sir James South's handwriting.)

The pension was granted on December 24.

On the 30th, Faraday writes to Lord Melbourne, to ask him to pass a box of scientific things through the Custom House. Lord Melbourne says in his answer, 'I beg to return you my thanks for your willingness, expressed through Dr. Holland, to contradict any injurious statements in the public prints. The best course was followed, which was to terminate the discussion as soon as possible.'

At the end of January, Faraday writes to Miss Fox, 'You will be glad to hear that his Lordship (Lord M.) expressed his approbation of the course pursued on a late occasion as being that which was decidedly the best.' She replies, 'You must allow me to say that in a case where an unfair impression has been given of the conduct of a man of such a character, he is the very last person to be consulted, or to require the contradiction of an assertion which he is conscious is utterly undeserved; but his friends probably did and do still think otherwise.'

In 1862, in the Life of Sir Charles Bell, an account like that in the 'Times,' of the interview between Lord Melbourne and Faraday, was about to be published.

1835.   The editor asked Faraday if he objected to the pub-
Æt. 43. lication. He answered—

TO B. BELL, ESQ.

'The Green, Hampton Court: August 27, 1862.

' Sir,—I am very much obliged to you for your
kindness in sending me the note regarding Lord Melbourne and my pension. I cannot assent to its publication. Lord Melbourne wrote to me and behaved
very handsomely in the matter ; and if one part of the
affair were published I think the other ought to be
also. But it was the desire of the Minister and his
friends that no notice should be taken of the affair, and
therefore I did not consent then (and cannot now) to
any partial publication of the circumstances.[1]

' I am, Sir, your very obedient servant,

' M. FARADAY.'

Another slight indication of his character and
opinions is seen in the following letter to his friend
Magrath. It is the only record that remains of a tour
in Switzerland which he made this year.

FARADAY TO MAGRATH.

' July 19, 1835.

' Dear Magrath,—What with occupation, fatigue,
and rheumatism, I have not yet been able to write to
you, but must redeem my promise at this place if I can,
or else I shall be at Albemarle Street, and then have no
power to do anything but listen to your reproaches. On
the whole we have done very well : what we have seen
we have enjoyed very much. We had a rough passage

---

[1] To avoid any partial publication, all the remaining letters regarding
the pension are given in the Appendix to this chapter. See p. 116.

to Dieppe from Brighton, so rough that we found the French people wondering that we had ventured, but were so unhappy in our sickness as to be quite unconscious of everything else. We passed through Rouen to Paris, and spent eight days at the latter place, took up Mr. Barnard there, and travelled through to Geneva ; there we stopped three or four days, and found De la Rive and our friends very kind ; and then left for Chamouni, where stopping two days we had time to go to the Montanvert, the Sea of Ice, the Flégère, the Glacier des Bossons, &c., and see all we wanted to see. We got to Martigni by the Tête Noire, and were quite satisfied with the choice of passage. From Martigni to Vevay and from Vevay to this place, where we are at present date, 19th. To-morrow we start for Berne, and then as things may turn up. Now you will see that we have not done and cannot do all we intended, but I do not know that that much matters.

' We are almost surfeited with magnificent scenery, and for myself I would rather not see it than see it with an exhausted appetite. The weather has been most delightful, and everything in our favour, so that the scenery has been in the most beautiful condition. Mont Blanc, above all, is wonderful, and I could not but feel at it what I have often felt before, that painting is very far beneath poetry in cases of high expression ; of which this is one. No artist should try to paint Mont Blanc ; it is utterly out of his reach. He cannot convey an idea of it, and a formal mass or a commonplace model conveys more intelligence, even with respect to the sublimity of the mountain, than his highest efforts can do. In fact, he must be able to dip his brush in light and darkness before he can paint

Mont Blanc. Yet the moment one sees it Lord Byron's
expressions come to mind, as they seem to apply. The
poetry and the subject dignify each other.

'There is a very fine iron wire bridge here above
900 feet in span and suspended at a great height over
the river and valley beneath. It is rough in finish, but
a fine work. We were on it last night, a fine scene of
lightning being all around us, and the effect was very
beautiful.

'We feel that we are now on our road home, and I
think we shall continue right on in our course and be at
home about August 1, but I cannot tell to a day or
two because of the uncertainty about steam-boats to
England, and because of the difficulty of choosing
either Holland or Belgium at pleasure as our route.
But I dare say that, thanks to your kindness, I shall find
help at Frankfort in Mr. Koch, as I did in Paris in
M. Feuillet, who desires his remembrances to you.
For the present farewell. Yours most truly,

'M. FARADAY.'

II.

The scientific work of the year is seen in his
papers, his notes, and his publications and lectures.
He sent a paper on an improved form of the voltaic
battery to the Royal Society. It formed the tenth
series of his 'Researches.' The experimental work for
this paper was chiefly done in the previous year. 'On
examining,' he says, 'however, what had been done
before, I found that the new trough was in all essential respects the same as that invented and described
by Dr. Hare, Professor in the University of Pennsylvania, to whom I have great pleasure in referring it.'

January 16th, he 'resumed the investigation of fluorine, &c.' On February 19 he 'arranged an electrolytic platinum retort for the decomposition of fluoride of lead.' But when in action there were no signs of free fluorine anywhere; and all February he worked with no result except finding that hot fluoride of lead conducts freely without decomposition.

In April he enters some queries in his note-book; one of these runs thus: ' If a freezing solution be well agitated during solidification—the ice ought to give pure water.'

On his return from Switzerland, August 6, his first thoughts are on electro-chemical decomposition.

On the 11th he writes, 'Very tired—cannot get energies up.'

On September 6 he says, 'Rose tells me that Berzelius in his annual account objects to my antimony proto-sulphuret, and I am therefore hastened to its examination the first thing this autumn, having meant to defer it awhile before Rose told me this.' More accurate experiments showed him his error, and he published a note on it in the ' Philosophical Magazine ' for June 1836.

By September 25 he is again at fluorine.

On November 3 he writes : ' Have been thinking much lately of the relation of common and voltaic electricity, of induction by the former and decomposition by the latter, and am quite convinced that there must be the closest connection. Will be first needful to make out the true character of ordinary electrical phenomena. The following notes are for experiment and consideration.

' Does common electricity reside upon the surface of

a conductor or upon the surface of the electric in contact with it?' And then he makes many pages of suggestions for experiments.

On November 6 he again works on fluorine ; on the
9th he says : 'It is scarcely likely that iron, nickel,
cobalt, and chrome, are the only magnetic bodies. It
is far more likely that certain temperatures are necessary for the effect, and that these bodies are on the
right side of the point of temperature required for
each, whilst many others are on the wrong side.'
And later in the month he makes more suggestions
for experiments on frictional electricity.

December 4th, he says, ' I cannot go on at present
with the fluorine experiments.'

The 5th, he began his experiments on frictional electricity, conduction, induction, transfer, &c., and these
he continued until Christmas Day. The 26th, he
experiments on the effect of low temperature on lead,
copper, gold, silver, platina, palladium, tin, cadmium,
zinc, plumbago, bismuth, antimony, arsenic, as regards
their magnetic power. The temperature was 60° or 70°
below zero.

At the beginning of the next year he went on with
his experiments on induction, &c.

He published in the ' Philosophical Magazine' a
reply to Dr. John Davy's remarks on certain statements
contained in the ' Researches.' And in the ' Proceedings'
of the Royal Society an account of the water of the
well Zem Zem.

At the Royal Institution, in May and June, he
gave a course of eight lectures on the chemical and
physical properties of the common metals, iron,
gold, platinum, lead, copper, zinc, mercury, tin, and
silver.

He gave the Juvenile Lectures on Electricity. The 1835.
course was the same as that in 1829. Æt.43-44.

He gave Friday discourses on Melloni's recent discoveries on radiant heat; on the Induction of Electric
currents ; on the manufacture of Pens from Quills and
Steel, illustrated by Mordan's machinery ; on the condition and use of the Tympanum of the Ear. He
ended his notes of this lecture with general remarks
on the relations of the powers of matter and the connection of the forces, &c.

This year he drew up notes for a new course of
fourteen lectures on electricity. These he gave as part
of Mr. Brande's morning course for the medical students
of St. George's Hospital. In his ninth lecture he said,
' It is most essential in the present state of science to
understand as accurately and distinctly as possible the
true relation existing between the chemical *generating*
force and the chemical *decomposing* and other forces of
the pile, first as connected with the question of *metallic contact*, next as to the *identity* of electrical and
chemical forces.' Later he says, ' All tends to prove
that chemical affinity and electricity are but different
names for the same power, and that all chemical phenomena are but exhibitions of electrical attractions.'

He ends thus : ' We know nothing of the intimate
nature of electricity, whether it is matter, force, vibration, or what. Of theories, there are several ; and two
principal ones, both assuming it to have an existence
independent of ordinary matter.'

III.

The marks of his reputation are to be seen in the
honours he received and in letters from Humboldt and

Melloni. In the previous year, Faraday had brought the great discoveries of Melloni on Radiant Heat before the Council of the Royal Society. The Rumford Medal was in consequence awarded to Melloni; his letters show not only his gratitude, but also the reputation which Faraday possessed.

This year he was made Corresponding Member of the Royal Academy of Medicine, Paris; Hon. Member of the Royal Society of Edinburgh, of the Institution of British Architects, and of the Physical Society of Frankfort; Hon. Fellow of the Medico-Chirurgical Society of London; and he was awarded one of the Royal Medals by the Royal Society.

BARON HUMBOLDT TO FARADAY.

' Potsdam: 11 janvier 1835.

' Monsieur,—Je suis bien, bien coupable, monsieur, d'avoir tardé si longtemps à vous remercier de l'aimable accueil que vous avez fait à ma prière relative aux gymnotes. Ce n'est pas autant la confusion jetée dans mes travaux par un voyage que j'ai dû faire avec le Roi à Königsberg-sur-mer, ni cette *conjonction planétaire* de tous les princes du nord dont Berlin a été le théâtre, ni un bras toujours bien malade depuis le séjour dans les forêts humides de l'Orénoque qui m'ont fait tarder; c'est l'espoir que j'avais de vous faire hommage, monsieur, de la première partie de mon *Cosmos* ("Physique du Monde," essai de présenter à la fois les phénomènes des cieux et de la géographie physique), qui de mois en mois a prolongé mon silence. Je crains bien que vous ne soyez pas initié dans le dédale de notre langue germanico-indico-pélagique; cependant, je tiens trop à l'es-

poir de placer dans votre bibliothèque le livre auquel
je mets le plus d'importance, parce qu'en peu de pages
il devrait offrir le plus de faits, pour que je résiste à la
tentation. Dès que la première partie sera imprimée,
j'aurai l'honneur de vous en adresser un exemplaire par
la voie du Baron de Bulow, ministre de Prusse à Londres, gendre de mon frère aîné. C'est par la même
légation que je vous offre aujourd'hui un beau groupe
de ces cristaux de feldspath qui ont pris naissance, pour
ainsi dire, sous nos yeux.

'Agréez, je vous supplie, l'hommage de ma vive
reconnaissance pour les démarches que vous avez faites
pour obtenir les gymnotes. Après les grandes et admirables découvertes que nous devons à votre sagacité,
monsieur, la Société Royale saura profiter des phénomènes étonnants qu'offre l'organisation animale dans sa
plus puissante action au dehors.

'Les gymnotes communs dans les Llanos de Caracas
(près de Calebaze), dans toutes les petites rivières qui
vont à l'Orénoque, dans les Guyanes anglaise, française,
et hollandaise, ne sont pas d'un transport difficile.
Nous les avons perdus sitôt à Paris parce qu'on les a
trop fatigués dès leur arrivée. A Stockholm, chez MM.
Norderling et Fahlberg, ils ont vécu quatre mois.
Je conseillerais de les transporter de Surinam, d'Esse-
quibo, Démérary, Cayenne, en été, car les gymnotes
vivent dans leur pays natal dans des eaux de 25°
Il y en a qui ont cinq pieds de long. Je conseillerais
d'en choisir de vingt-sept à vingt-huit pouces. Leur
force varie avec la nourriture et le repos. Ils man-
gent, ayant l'estomac très-petit, peu et souvent, de la
viande cuite non salée, de petits poissons, même du
pain. Il faut essayer leur force et le genre de nour-

riture avant de les embarquer, et ne prendre que des poissons déjà accoutumés à la prison. Je les ai tenus dans un baquet long de quatre pieds et de seize pouces de large. Il faut changer l'eau (douce) tous les trois ou quatre jours, et ne pas empêcher le poisson de venir à la surface, car il aime à sentir l'air. Un filet doit entourer le baquet. Le gymnote s'élance volontiers hors de l'eau. Voilà tous les conseils que je saurais donner. Il est important surtout de ne pas trop tourmenter l'animal : il s'épuise par les fréquentes explosions électriques. Plusieurs gymnotes peuvent être dans un même baquet. Je ne connais de M. Davy que l'intéressant mémoire sur la torpille et sa faculté de décomposer l'eau (" Phil. Trans." 1832, Ptie. II. p. 259). Nous ne possédons encore ici que la première partie des "Trans." de 1834, qui renferment vos importants mémoires du 11 janvier et du 13 février 1834. Quoique mes propres travaux de la chimie pratique tombent dans le monde antédiluvien, je n'en suis pas moins attentif aux progrès d'une science à laquelle je dois les plus douces jouissances intellectuelles. Si encore, l'hiver passé, j'ai assisté tous les matins aux cours de M. Mitscherlich, qui professe de la manière la plus distinguée, c'était surtout pour voir une partie de ces belles expériences que nous devons à votre sagacité, monsieur. De nouvelles manières de voir nécessitent sans doute une nouvelle nomenclature. Ces grands mots de positif et de négatif ont fait bien du mal, presque autant que les abus des mots froid et chaud en médecine, chirurgie et matière médicale. Votre nomenclature s'adapte au génie des langues de l'Europe latine. Les Français ne se révoltèrent que contre les " anions " et les " ions." Une nomenclature est toujours bonne lorsque, comme la vôtre, elle étend la sphère de nos idées.

'Agréez, je vous supplie, monsieur, l'expression de la haute et affectueuse admiration qui vous est due à tous les titres.

'Votre très-humble et très-obéissant serviteur,
'ALEXANDRE HUMBOLDT.

'Je désire bien que M. Davy ait traité la question de l'action de la torpille dans les cas où l'on ne forme pas de chaîne. Je fais allusion à des expériences que j'ai faites avec M. Gay-Lussac, et qui ne sont bien développées que dans la "Rélat. hist." de mon voyage (tom. ii. chap. vii. p. 186). Dans l'état actuel de la science il sera facile à cet excellent physicien de trancher la question vitale sur la mode d'action. Je n'ai malheureusement su que pour quelques heures votre mémoire imprimé dans le "Journal of Science" (novembre, p. 334). Ce que vous exposez sur " *the definite nature* " des décompositions électro-chimiques est d'une haute portée. C'est une grande loi de la nature.'

M. MACÉDOINE MELLONI TO FARADAY.

'Paris : le 4 fév. 1835.

'Monsieur,—Il y a dans la vie certains événements qui causent une sensation trop vive pour pouvoir s'exprimer d'une manière convenable. Tel est l'effet qu'a produit sur moi votre noble conduite dans l'affaire de la Société Royale. Ici je fréquentais des académiciens puissants qui ne cessaient de me prodiguer des paroles dorées ; parmi eux il s'en trouvait quelques-uns auxquels j'avais rendu des services, et plusieurs qui se disaient mes amis. Ces messieurs voyaient les obstacles

que la malignité opposait à ma carrière scientifique.
Ils pouvaient les écarter—je dirai plus, ils le devaient ;
car c'était au jugement du corps savant dont ils font
partie que je soumettais le fruit de mes veillées ; et,
cependant, la justice que j'avais le droit d'attendre d'eux
était indéfiniment prorogée dans la seule crainte d'user
leur influence auprès des méchants. On ne pouvait
nier en public des faits évidents et connus par la grande
masse des philosophes indépendants ; il fallait donc
tâcher de les faire oublier par un silence officiel.
On risquait ainsi d'étouffer des germes qui pouvaient
devenir féconds pour la science ; mais qu'importe?
Périsse la science et la justice plutôt que nos intérêts!
Voilà leur devise.

' Et vous, monsieur, qui appartenez à une société à
laquelle je n'avais rien offert, vous, qui me connaissiez
à peine de nom, vous n'avez pas demandé si j'avais
des ennemis faibles ou puissants, ni calculé quel en était
le nombre ; mais vous avez parlé pour l'opprimé étran-
ger, pour celui qui n'avait pas le moindre droit à tant
de bienveillance, et vos paroles ont été accueillies favora-
blement par des collègues consciencieux ! Je reconnais
bien là des hommes dignes de leur noble mission, les
véritables représentants de la science d'un pays libre et
généreux. Ailleurs tout est égoïsme ou déception.

' Que mille et mille grâces soient rendues en mon
nom à M. Faraday et au Conseil de la Société Royale.
Je n'y saurais ajouter autre chose pour le moment, mais
j'attends avec impatience l'occasion de montrer par des
faits les sentiments ineffaçables de reconnaissance qui
sont profondément gravés dans le cœur de
    ' Votre tout-dévoué serviteur et ami,
        ' MACÉDOINE MELLONI.'

M. MACÉDOINE MELLONI TO FARADAY.

'Paris: le 6 mars 1835.

'J'apprends par votre dernière que l'on vous a consigné la médaille de Rumford, et que vous voudriez savoir quelle serait la voie la plus convenable de me la faire parvenir.

. . . . . .

'Quant à moi, monsieur, je ne puis que vous remercier de la peine que vous vous êtes donnée de répéter mes dernières expériences à l'Institution Royale ; et, puisque vous prenez tant d'intérêt dans ce qui me regarde, je me fais un véritable plaisir de vous annoncer que MM. Biot, Poisson et Arago ont examiné mes résultats dans le plus grand détail, qu'ils en sont *enthousiasmés*, et que le premier va bientôt en faire en leur nom un rapport extrêmement favorable à l'Institut. Pour cette fois la chose est certaine ; déjà, comme arrhes, le ministre d'Instruction publique m'a accordé une somme de 1200 francs à leur sollicitation. . . . Je dois de la reconnaissance à ces messieurs, et j'en aurais ; mais mes compatriotes observent que tout cela arrive après le prix de la Société Royale; et moi, je ne puis m'empêcher de réfléchir que la Société Royale n'aurait pas pensé à me décerner cette récompense sans vos soins et votre amitié. . . . Vous voyez donc que je vous dois tout ; aussi mon cœur est-il profondément ému toutes les fois que je me déclare

'Votre très-dévoué et très-reconnaissant

'MACÉDOINE MELLONI.'

In 1836, 1837, and 1838, the life of Faraday is to be seen chiefly in his work. He was fully occupied

1836–38. in his electrical researches. He did much also for the
Æt.44–47. Royal Institution by his lectures; and he undertook
new work as scientific adviser to the Trinity House. He
received many honours, and his correspondence shows
some of the kindness and nobleness of his character.

I.

His laboratory notes show the mass of experiments
which he made during 1836 and the following year on
the induction, conduction, and discharge of common
electricity, and its relation to voltaic electricity. Some
of his entries at this time give an insight into the
working of his mind.

February 1, he writes : ' Is evident in voltaic battery
with its *tension* and its *spark* that *chemical action is
electricity.* Hence also *electricity is chemical action.*
Hence electricity of rubbed glass should be chemical
action. Hence chemical tension of *acid* and *amalgamated zinc,* before the current passes, must be the same
as rubbing glass and amalgam, or rubbing wax and
flannel, *in its origin.* Induction of battery poles or
electrodes in an experimental decomposing cell, or in
the air with a spark, must be the same force. Hence
glass electricity and its induction up to the spark must
be the same force.' &c.

On July 28, under Thermo-Electricity, he writes:
' Surely the converse of thermo-electricity ought to be
obtained experimentally. Pass current through a
circuit of antimony and bismuth or through the compound instrument of Melloni.'

On August 3 he writes : ' After much consideration
(here at Ryde) of the manner in which the electric
forces are arranged in the various phenomena gene-

rally, I have come to certain conclusions which I will 1837.
endeavour to note down without committing myself to Æт.45–46.
any opinion as to the cause of electricity, i.e. as to the
nature of the power. If electricity exist independently
of matter, then I think that the hypothesis of one
fluid will not stand against that of two fluids. There
are, I think, evidently, what I may call two elements of
power of equal force and acting towards each other.
These may conventionally be represented by oxygen
and hydrogen, which represent them in the voltaic
battery. But these powers may be distinguished only
by direction, and may be no more separate than the
north and south forces in the elements of a magnetic
needle. They may be the polar points of the forces
originally placed in the particles of matter; and the
description of the current as an axis of power which
I have formerly given suggests some similar general
impression for the forces of quiescent electricity. Law
of electric tension might do, and though I shall use the
terms positive and negative, by them I merely mean
the termini of such lines.'

Throughout the whole of 1837 the experimental
researches were continued, and on November 30 he
sent the result of upwards of two years of hard work
to the Royal Society. It forms the eleventh series.
The twelfth series on the same subject he sent on
January 11, the thirteenth on February 22, and the
fourteenth on June 21, 1838.

Dr. Tyndall gives the following admirable summary
of these researches :—

' His first great paper on frictional electricity was
sent to the Royal Society on November 30, 1837. We
here find him face to face with an idea which beset his

1837.
Æt.45–46.

mind throughout his whole subsequent life,—the idea of *action at a distance.* It perplexed and bewildered him. In his attempts to get rid of this perplexity he was often unconsciously rebelling against the limitations of the intellect itself. He loved to quote Newton upon this point : over and over again he introduces his memorable words, " That gravity should be innate, inherent, and essential to matter, so that one body may act upon another at a distance through a *vacuum* and without the mediation of anything else, by and through which this action and force may be conveyed from one to another, is to me so great an absurdity, that I believe no man who has in philosophical matters a competent faculty of thinking can ever fall into it. Gravity must be caused by an agent acting constantly according to certain laws ; but whether this agent be material or immaterial I have left to the consideration of my readers." [1]

' Faraday does not see the same difficulty in his contiguous particles. And yet by transferring the conception from masses to particles we simply lessen size and distance, but we do not alter the quality of the conception. Whatever difficulty the mind experiences in conceiving of action at sensible distances, besets it also when it attempts to conceive of action at insensible distances. Still the investigation of the point whether electric and magnetic effects were wrought out through the intervention of contiguous particles or not, had a physical interest altogether apart from the metaphysical difficulty. Faraday grapples with the subject experimentally. By simple intuition he sees that action at a distance must be exerted in straight lines. Gravity,

---

[1] Newton's third letter to Bentley.

he knows, will not turn a corner, but exerts its pull 1837.
along a right line ; hence his aim and effort to ascertain Æt.45-46.
whether electric action ever takes place in curved lines.
This once proved, it would follow that the action is
carried on *by means of a medium* surrounding the
electrified bodies. His experiments in 1837 reduced,
in his opinion, this point to demonstration. He then
found that he could electrify by induction an insulated
sphere placed completely in the shadow of a body
which screened it from direct action. He pictured the
lines of electric force bending round the edges of the
screen, and reuniting on the other side of it ; and he
proved that in many cases the augmentation of the
distance between his insulated sphere and the inducing
body, instead of lessening, increased the charge of the
sphere. This he ascribed to the coalescence of the
lines of electric force at some distance behind the
screen.

'Faraday's theoretic views on this subject have not
received general acceptance, but they drove him to
experiment, and experiment with him was always
prolific of results. By suitable arrangements he places
a metallic sphere in the middle of a large hollow
sphere, leaving a space of something more than half
an inch between them. The interior sphere was insulated, the external one uninsulated. To the former
he communicated a definite charge of electricity. It
acted by induction upon the concave surface of the
latter, and he examined how this act of induction was
affected by placing insulators of various kinds between
the two spheres. He tried gases, liquids, and solids,
but the solids alone gave him positive results. He
constructed two instruments of the foregoing descrip-

1837. tion, equal in size and similar in form. The interior
Æт.45-46. sphere of each communicated with the external air by
a brass stem ending in a knob. The apparatus was
virtually a Leyden jar, the two coatings of which were
the two spheres, with a thick and variable insulator
between them. The amount of charge in each jar
was determined by bringing a proof-plane into contact
with its knob, and measuring by a torsion balance the
charge taken away. He first charged one of his instruments, and then dividing the charge with the other,
found that when air intervened in both cases, the charge
was equally divided. But when shell-lac, sulphur, or
spermaceti was interposed between the two spheres of
one jar, while air occupied this interval in the other,
then he found that the instrument occupied by the
"solid dielectric" took *more than half* the original
charge. A portion of the charge was absorbed in the
dielectric itself. The electricity took time to penetrate
the dielectric. Immediately after the discharge of the
apparatus no trace of electricity was found upon its
knob. But after a time electricity was found there,
the charge having gradually returned from the dielectric in which it had been lodged. Different insulators
possess this power of permitting the charge to enter
them in different degrees. Faraday figured their particles as polarised, and he concluded that the force
of induction is propagated from particle to particle of
the dielectric from the inner sphere to the outer one.
This power of propagation possessed by insulators he
calls their "*Specific Inductive Capacity.*"

'Faraday visualises with the utmost clearness the
state of his contiguous particles; one after another
they become charged, each succeeding particle depend-

ing for its charge upon its predecessor. And now he
seeks to break down the wall of partition between
conductors and insulators. " Can we not," he says, " by
a gradual chain of association carry up discharge from
its occurrence in air through spermaceti and water to
solutions, and then on to chlorides, oxides, and metals,
without any essential change in its character ? " Even
copper, he urges, offers a resistance to the transmission
of electricity. The action of its particles differs from
those of an insulator only in degree. They are
charged like the particles of the insulator, but they
discharge with greater ease and rapidity; and this
rapidity of molecular discharge is what we call conduction. Conduction then is always preceded by
atomic induction; and when through some quality of
the body, which Faraday does not define, the atomic
discharge is rendered slow and difficult, conduction
passes into insulation.

' Though they are often obscure, a fine vein of philosophic thought runs through those investigations. The
mind of the philosopher dwells amid those agencies
which underlie the visible phenomena of induction and
conduction; and he tries by the strong light of his
imagination to see the very molecules of his dielectrics.
It would, however, be easy to criticise these researches,
easy to show the looseness, and sometimes the inaccuracy,
of the phraseology employed; but this critical spirit
will get little good out of Faraday. Rather let those
who ponder his works seek to realise the object he set
before him, not permitting his occasional vagueness to
interfere with their appreciation of his speculations.
We may see the ripples, and eddies, and vortices of a

flowing stream, without being able to resolve all these
motions into their constituent elements; and so it
sometimes strikes me that Faraday clearly saw the
play of fluids and ethers and atoms, though his previous
training did not enable him to resolve what he saw
into its constituents, or describe it in a manner satisfactory to a mind versed in mechanics. And then again
occur, I confess, dark sayings, difficult to be understood,
which disturb my confidence in this conclusion. It
must, however, always be remembered that he works
at the very boundaries of our knowledge, and that his
mind habitually dwells in the " boundless contiguity of
shade " by which that knowledge is surrounded.

' In the researches now under review the ratio of
speculation and reasoning to experiment is far higher
than in any of Faraday's previous works. Amid much
that is entangled and dark we have flashes of wondrous
insight and utterances which seem less the product of
reasoning than of revelation. I will confine myself
here to one example of this divining power :—By his
most ingenious device of a rapidly rotating mirror,
Wheatstone had proved that electricity required time
to pass through a wire, the current reaching the middle
of the wire later than its two ends. " If," says Faraday,
" the two ends of the wire in Professor Wheatstone's
experiments were immediately connected with two
large insulated metallic surfaces exposed to the air, so
that the primary act of induction, after making the
contact for discharge, might be in part removed from
the internal portion of the wire at the first instance,
and disposed for the moment on its surface jointly with
the air and surrounding conductors, then I venture to

anticipate that the middle spark would be more retarded
than before. And if those two plates were the inner
and outer coatings of a large jar or Leyden battery,
then the retardation of the spark would be much
greater." This was only a *prediction*, for the experiment was not made. Sixteen years subsequently,
however, the proper conditions came into play, and
Faraday was able to show that the observations of
Werner Siemens and Latimer Clark on subterraneous
and submarine wires were illustrations, on a grand
scale, of the principle which he had enunciated in 1838.
The wires and the surrounding water act as a Leyden
jar, and the retardation of the current predicted by
Faraday manifests itself in every message sent by such
cables.

' The meaning of Faraday in these memoirs on induction and conduction is, as I have said, by no means
always clear ; and the difficulty will be most felt by
those who are best trained in ordinary theoretic conceptions. He does not know the reader's needs, and
he therefore does not meet them. For instance, he
speaks over and over again of the impossibility of
charging a body with one electricity, though the impossibility is by no means evident. The key to the
difficulty is this. He looks upon every insulated conductor as the inner coating of a Leyden jar. An insulated sphere in the middle of a room is to his mind
such a coating ; the walls are the outer coating, while
the air between both is the insulator, across which the
charge acts by induction. Without this reaction of
the walls upon the sphere, you could no more, according
to Faraday, charge it with electricity than you could

charge a Leyden jar, if its outer coating were removed. Distance with him is immaterial. His strength as a generaliser enables him to dissolve the idea of magnitude; and if you abolish the walls of the room—even the earth itself—he would make the sun and planets the outer coating of his jar. I dare not contend that Faraday in these memoirs made all these theoretic positions good. But a pure vein of philosophy runs through these writings; while his experiments and reasonings on the forms and phenomena of electrical discharge are of imperishable importance.'

Two notes for future work—taken from his laboratory book—must be mentioned here. August 24, he writes: 'With respect to crystallisation, endeavour to carry crystallising influence by wires, so as to set off, as it were, another solution beside the active one, i.e. try to transfer the crystallising force.' And November 14, he writes : ' Compare corpuscular forces in their amount, i.e. the forces of electricity, gravity, chemical affinity, cohesion, &c., and give, if I can, expressions of their equivalents in some shape or other.'

In 1838, in addition to the twelfth, thirteenth, and fourteenth series of ' Researches,' the fifteenth was sent in November to the Royal Society ; this was on the character and direction of the electric force of the gymnotus. The fish ' was brought to this country about four weeks ago,' Faraday says in his notes of September 3 ; ' it is about three feet six inches in length ; has not fed whilst here or on the passage to this country, and is probably very languid.' The experiments were continued in October and November at the Adelaide Gallery. The gymnotus thrived on fish. In addition

to the shock it showed by the galvanometer a current
outside the fish, from the head to the tail, capable of
making a magnet, of causing chemical decomposition,
and of giving a spark. He estimated the medium discharge as at least equal to the electricity of a Leyden
battery of fifteen jars, containing 3,500 square inches
of glass, coated on both sides, charged to its highest
degree.

At the end of the year he was again experimenting
on the source of power in the voltaic pile.

In addition to his researches in electricity he did
other experimental work in these three years.

In January 1836 he worked upon the action of
heat as regards its influence on the magnetism of iron.
This, and his work in the previous year, he published
in a paper in the 'Philosophical Magazine,' on the
general magnetic relations and characters of the metals.
He begins by saying, 'General views have long since
led me to an opinion, which is probably also entertained by others, though I do not remember to have
met with it, that *all* the metals are magnetic in the
same manner as iron.'

He had another paper in the same Magazine, on a
supposed new sulphate and oxide of mercury; and
another on the history of the condensation of gases;
and another on a peculiar voltaic condition of iron.

In 1837, in the same Magazine, he had a paper on
the causes of the neutrality of iron in nitric acid.

His work for the Royal Institution in 1836 was six
lectures, after Easter, on the Philosophy of Heat.

He also gave four Friday discourses—on Silicified
Plants and Fossils; on Magnetism of Metals as a gene-

1837-38. ral character ; on Plumbago, and on Pencils, Mordan's
Æᴛ.44-47. Machinery ; and considerations respecting the nature
of Chemical Elements. He thus ended this lecture on
June 10 :

' Thus either present elements are the true elements,
or else there is the probability before us of obtaining
some *more high and general power* of nature even
than electricity ; which at the same time might reveal
to us an entirely *new* grade of the elements of matter,
now hidden from our view, and almost from our sus-
picion. This is the high prize set before the chemico-
physical philosopher of the present day by the present
state of our scientific knowledge, and with the pro-
pounding of it I have thought I might well conclude
this series of Friday evenings.'

In 1837, at the Institution, he gave six lectures after
Easter, on Earth, Air, Fire, and Water.

Speaking of silica he says, ' It startles us by the
strange places in which we find it. These things un-
accountable at present, but show us that with all our
knowledge, we know little as yet of that which *may* be
known.'

His four Friday discourses were on the views of
Professor Mossotti as to one general law accounting for
the different forces in matter. He ends his notes with :
' Mossotti's words not impossible ; want experimental
proof of the general law. Progress of knowledge not
in floods—dangerous as floods of water, but a calm
and dignified process. *Nature of a thing ;* the answer
both of the ignorant and the philosopher. Search for
laws.' On Dr. Marshall Hall's views of the Nervous
System ; on De la Rue's mode of applying sulphate of

copper to the exaltation of the power of a common 1836–38. voltaic battery ; on the peculiar relation of iron.   In 1838, after Easter, at the Institution, he gave eight lectures on Electricity. He ends his notes of the last lecture on June 17, thus : ' General experimental relation of the powers ; ignite wire ; give spark ; decompose ; make magnet by one current, and hence universality of the common cause, whatever it may be ; the force is nowhere destroyed ; all the effects are convertible.' He gave three Friday discourses this year on Mr. Ward's plan of growing plants in cases. May 18, on the gaseous fluid and solid condition of carbonic acid (' Not impossible that hydrogen a metal ') ; on insulation and conduction.

For the Trinity House, in 1836, his first work was to make a photometer. Throughout the whole year he was busy on the subject, making three photometers, and ascertaining the capability and accuracy of the instruments. He also experimented on the preparation of oxygen for the Bude light, drawing up the most exact tables for the record of the manufacture ; for example, on November 10, he says, 'Hence oxygen costs very nearly twopence per cubical foot ; exactly 1·909 pence.'

In 1837 his work for the Trinity House consisted in examining the Trinity lamp, the French lamp, and the Bude lamp, as to intensity of light and price : ' pressed Mr. Gurney, by letter, to give us his best lamp at once, and not to lose time.'

In 1838, for the Trinity House, he a second time reported on the new Gurney lamp, comparing it in light and cost with the French lamp.

1836–38.   Addresses were made to Lord John Russell for the
Æt.44–47. removal of the cartoons to the National Gallery ; and
Faraday was asked if it was likely to be prejudicial to
them.

FARADAY TO J. PHILLIPPS, ESQ.

(UNDER-SECRETARY OF STATE FOR THE HOME DEPARTMENT.)

'Royal Institution : July 19, 1838.

' Sir,—I am in town for a day only, but hasten to
answer your letter. I feel much diffidence in forming
an opinion respecting the effect of the air of the metropolis upon the cartoons without previous consultation
with those who have been conversant with such action ;
but to avoid delay will state at present that I should
not anticipate any harm from the chemical action of
the air, either upon the colours or upon the vehicle or
medium by which they are applied.

' But there is another effect of the London atmosphere
brought to my mind by Mr. Phillips, of the Royal
Academy, from which I should fear much harm ; I
mean the dirtying effect. The ceilings and walls of a
London room sufficiently show to what extent dirt in
the form of dust may *penetrate* surfaces and textures
like those of the distemper-painted cartoons. In oil
paintings, the dust, if it adheres, is stopped upon the
surface, and the substances applied for its removal need
not pass through the varnish ; so that the colours themselves are not necessarily at any time either injured by,
or in contact with, the dirtying or the cleansing medium.
But in such paintings as the cartoons, the dust has
access to the very colours and body of the picture, and

then cannot be dislodged without causing the destruc- 1836–38.
tion, or something very like it, of the whole.   Æt.46–47.

'As I see no effectual way of preventing the access
of dirt to the pictures, I fear that the evil thus arising
may, in the metropolis, be very considerable, and
produce an effect in the lapse of thirty, forty, or fifty
years which could never be remedied.

'I have the honour to be, Sir, your obedient, humble
servant,

'M. FARADAY.'

II.

In 1836 he was made Senator of the University of
London; Hon. Member of the Society of Pharmacy of
Lisbon and of the Sussex Royal Institution; Foreign
Member of the Society of Sciences of Modena, and the
Natural History Society of Basle.

In 1837 he was elected Honorary Member of the
Literary and Scientific Institution, Liverpool.

In 1838 he was made Honorary Member of the
Institution of Civil Engineers; Foreign Member of the
Royal Academy of Sciences, Stockholm; and he re-
ceived the Copley Medal.

III.

Kindness and consideration for others were the characteristics of all the most public and private acts of
Faraday's life, and these come out in strange ways

during these three years. It seems almost impossible that these qualities could appear in business with the Athenæum Club and the Trinity House. His brother was a gas-fitter, and Faraday writes to the Secretary: ' Few things would please me more than to help my brother in his business, or than to know that he had got the Athenæum work ; but I am exceedingly jealous of myself, lest I should endeavour to have that done for him as my brother which the Committee might not like to do for him as a tradesman, and it is this which makes me very shy of saying a word about the matter.'

His letter to the Deputy-Master of the Trinity House regarding his appointment as scientific adviser tells of his indifference to the proposal as a matter of interest, but *not as a matter of kindness.*

On February 3, 1836, he wrote to Captain Pelly, Deputy-Master of the Trinity House :—

' I consider your letter to me as a great compliment, and should view the appointment at the Trinity House, which you propose, in the same light ; but I may not accept even honours without due consideration.

' In the first place, my time is of great value to me, and if the appointment you speak of involved anything like periodical routine attendances, I do not think I could accept it. But if it meant that in consultation, in the examination of proposed plans and experiments, in trials, &c., made as my convenience would allow, and with an honest sense of a duty to be performed, then

I think it would consist with my present engagements.
You have left the title and the sum in pencil. These
I look at mainly as regards the character of the
appointment; you will believe me to be sincere in
this, when you remember my indifference to your
proposition as a matter of interest, though *not as a
matter of kindness.*

'In consequence of the goodwill and confidence of
all around me, I can at any moment convert my time
into money, but I do not require more of the latter
than is sufficient for necessary purposes. The sum,
therefore, of 200*l.* is quite enough in itself, but not if
it is to be the indicator of the character of the appointment; but I think you do not view it so, and that you
and I understand each other in that respect; and your
letter confirms me in that opinion. The position which
I presume you would wish me to hold is analogous to
that of a standing counsel.

'As to the title, it might be what you pleased almost.
Chemical adviser is too narrow; for you would find
me venturing into parts of the philosophy of light not
chemical. Scientific adviser you may think too broad
(or in me too presumptuous); and so it would be, if by
it was understood all science. It was the character I
held with two other persons at the Admiralty Board in
its former constitution.

'The thought occurs to me whether, after all, you
want such a person as myself. This you must judge
of; but I always entertain a fear of taking an office in
which I may be of no use to those who engage me.
Your applications are, however, so practical, and often
so chemical, that I have no great doubt in the matter.'

1837.   For thirty years nearly he held this post. What he
Æт.45-46. did may be seen in the portfolios, full of manuscripts,
which Mrs. Faraday has given to the Trinity House, in
which, by the marvellous order and method of his
notes and indices, each particle of his work can be
found and consulted immediately.
The following letter to M. C. Matteucci shows how
sound his judgment was, even in his own cause.

FARADAY TO MATTEUCCI.

'Royal Institution: April 19, 1839.

'My dear Sir,—On returning to town I find your
letter.[1] It rather embarrasses me as to the right mode
of proceeding, as it calls on me to publish that letter.
Now I have no doubt you were unacquainted with the
seventh series of my " Researches " when your paper was
written, and you probably had not recollected the an-

---

[1] On March 12, M. M. wrote to say that Poggendorf's Journal had
brought against him (M. M.) an unjust accusation of plagiarism from F.
He continued: 'In the Institut, Paris, October 18, 1834, the first notice
is found of your admirable labours, and it was before that epoch that the
memoir directed to M. Gay-Lussac was published, as the date will testify.
. . . . I flatter myself that you will have the goodness to give to this
letter the publicity which is required by consideration for my character
and by the high esteem and sincere friendship that I entertain towards
yourself. Believe me, &c., C. Matteucci.'
June 11, 1836, M. M. writes : 'Personne, monsieur, n'est plus que moi
convaincu de la date antérieure de votre grande découverte sur la force
décomposante du courant électrique, et vous pouvez regarder cette lettre
comme la déclaration la plus solennelle. . . . . . .
'Je ne connais M. Poggendorf, et je ne connais de quelle manière lui
passer cette déclaration.'
Then he asks F. to get him named Professor of Physics or Chemistry at
Corfu.

nouncement of the law in the third series, dated as far back as December 1832, *that the chemical power of a current of electricity is in direct proportion to the quantity of electricity which passes.* This law was again announced in the fifth series of my " Researches," of the date of *June* 1833, see paragraphs 456, 504, 505. I think you have these papers and can refer to them.

' My difficulty is this : I had not noticed the occurrence of your paper of a date so much later than my own, though many persons had pointed it out to me in the " Annales de Chimie," and expressed their surprise at it. It certainly had in their minds made an impression.

' The editor of " Poggendorf's Journal " did it of his own free will and judgment, and I think the plain course is for you to write to him, telling him that you were not acquainted with my paper ; but now that you do know the facts, acknowledge the order of the dates as they really stand.

' This is what I have always done, and you will see a case of it where I have made restitution to a countryman of your own, M. Bellani, in the " Quarterly Journal of Science," published here formerly, volume xxiv. pp. 469, 470.

' If you still wish me to publish the letter which you sent me, I hope you will write to me at once by post ; but I must accompany it with dates. I should, however, in my own case, pursue the plan I have recommended to you.

' I am, dear Sir, yours very faithfully,
' M. FARADAY.

FARADAY TO DR. BOOTH

(in reply to a request for a testimonial to his friend R. Phillips, who was candidate for the chair of Chemistry at University College).

'March 1, 1837.

'My dear Sir,—That I cannot give a testimonial to my friend Phillips is not merely a matter of general reluctance to certify, but of principle ; for refusing all, I am obliged to refuse each ; and I refuse all because, having no confidence in certificates, I wish to have no association in any way with them. In the present particular case, too, if I certify for one, I should have to certify for two others also, one being our friend Graham, of Glasgow. But when I am asked by an authority (concerned in forming the decision) what I think of a candidate, I cannot refuse to answer.

'To all that you have said of Phillips I fully agree. I should indeed have thought his character had been known to be such that it would rather have been degraded than established by certificates. What he has done in connection with the Pharmacopeia is fully sufficient to show the confidence reposed in his talents and abilities by those who hold high station in the medical world, and I can freely say that if I had time and desire to pursue still further pharmaceutical chemistry, I should go to Mr. Phillips for my teacher.

'I am, my dear Sir, very faithfully yours,

'M. FARADAY.'

1837.

_Æт. 46._

FARADAY TO MRS. FARADAY.

'British Association. Liverpool, Toxteth Park:
Tuesday, September 13, 1837.

. . . . . .

' We arrived here very safely and comfortably on
Monday about four o'clock, and since then have been
in a continual hurry, but our hosts here are in still
greater haste, and indeed labour too much in the
hospitable cause. Mr. Currie was waiting for us.

'This morning I went into town early after breakfast, and met Daniell with our infinity of friends. But
what will you think of Daniell's management, when I
tell you that besides separating us, as it has done, in
our domiciles, it has made us most responsible persons,
for I am President of the Chemical Section, and _he is
one of the three Vice-Presidents?_ Only think of our not
working. Why, it could not be. For if, after the
extreme kindness and forbearance which the friends
here showed to us, we had refused altogether to
join in the common feeling, we should have looked
like churls indeed. So we are in harness a little;
nothing like what we might be : for all are excessively
kind.

' To-day I think we made our section rather more
interesting than was expected, and to-morrow I expect
will be good also. In the afternoon Daniell and I took
a quiet walk ; in the evening he dined with me here.
We have been since to a grand conversazione at the
town-hall, and I have now returned to my room to talk
with you, as the pleasantest and happiest thing I can do.

Nothing rests me so much as communion with you. I feel it even now as I write, and I catch myself saying the words aloud as I write them, as if you were within hearing.

.   .   .   .   .   .

‘ Ever, my dear Sarah, your affectionate husband,

‘ M. FARADAY.’

This year the feeling of Faraday regarding his pension is again made clear by a letter to the Chancellor of the Exchequer, who was about to move for a parliamentary committee, regarding the continuance of pensions.

FARADAY TO THE RIGHT HON. T. SPRING RICE.

‘ Royal Institution : December 1, 1837.

‘ Sir,—I am honoured and much obliged by your communication, but have little to say in reply. When Lord Melbourne favoured me with a letter on the subject he was pleased to say that “ the distinction so bestowed ” was with the “ desire to reward acknowledged merit, and to advance the interests of science.” I am no judge whether that in the present instance be the case or not ; but if the grant do not retain the same feeling and character as that which his Lordship attached to it, I should, though with all respect to the Government, certainly have no wish for its continuance.

‘ I have the honour to be, Sir, your very obedient, humble servant,

‘ M. FARADAY.’

April 17, 1838, he received a circular from Mr. 1838.
Spring Rice, saying that he was at present engaged in Æt. 46.
preparing materials for a report on civil-list pensions,
and asking for a list of his works and titles. Faraday
replies:

FARADAY TO THE RIGHT HON. T. SPRING RICE.

'Royal Institution: April 23, 1838.

' Sir,—Though unwilling to do anything which might
bear the interpretation of a desire on my part to proffer
evidence in favour of my own character, I cannot for
a moment hesitate to answer your inquiries. Had I
been in town, the list I now enclose would have been
sent sooner. One title, namely that of F.R.S., was
sought and paid for; all the rest are spontaneous
offerings of kindness and goodwill from the bodies
named. Perhaps I ought to have added one more, for
I feel it an honour equal to that of any of those set
down—I mean that of *Member of the Senate of the
University of London*; but as you, Sir, were the person
conferring it, I have left it for you to do with as you
may please.

'I have the honour to be, Sir, your obedient, humble
servant,

' M. FARADAY.

' Corresponding Member Académie Royale des
Sciences, Institut de France; Société de Chimie Médi-
cale de Paris; Société Philomathique de Paris; Royal
Academy of Sciences, Berlin; Academy of Science and
Belles-Lettres, Palermo; Royal Academy of Medicine,

Paris; Physical Society of Frankfort; Natural History
Society of Basle.

'Foreign Member of Philadelphia College of Pharmacy; Royal Society of Göttingen; Modena Society
of Science.

'Fellow of the Society for Natural Sciences, Heidelberg; American Academy Arts and Sciences, Boston.

'Member of Royal Society of Sciences at Copenhagen.

'D.C.L. of Oxford University.

'F.R.S.

'Fullerian Professor of Chemistry, R.I.

'In " Phil. Trans.," nine papers on different subjects;
thirteen papers on Electricity—twenty-two.

'Many long and short in " Quarterly Journal of
Science," Chemical Manipulation.

'Honorary Member of Cambridge Philosophical
Society; Bristol Institution; Cambrian Society, Swansea; Society of Arts for Scotland; Imperial Academy
of Sciences, St. Petersburg; Society of Physical and
Chemical Sciences at Paris; Hull Literary and Philosophical Society; Institute of British Architects; Royal
Society of Edinburgh; Medical and Chirurgical Society
of London; Society of Pharmacy of Lisbon.

'Corresponding Associate of Imperial and Royal
Academy dei Georgofili di Firenze.'

In 1839 and 1840 much less original research was
done by Faraday than had been done in the previous
years. He began to feel the effects of overwork. Still
he added two more papers to the series of ' Experimental Researches in Electricity,' and he went on with
his work for the Institution, and for the Trinity House.

The most remarkable event of his life in 1840 was his election as an elder by the Sandemanian Church ; he held the office only for three years and a half. During that period when in London he preached on alternate Sundays. This was not entirely a new duty. From the time of his admission into the Church he had been occasionally called upon by the elders to exhort the brethren at the week-day meetings ; now, however, it was done regularly, and how thoroughly, Faraday's character, as seen up to this time, is sufficient to show. Certainly no more rest would be given by this new duty to his overworked mind.

It is very difficult to draw a comparison between his preaching and his lecturing : first, because they were very unequally known ; and secondly, because of the entire separation he made between the subjects of religion and of science.

Generally perhaps it might be said that no one could lecture like Faraday, but that many might preach with more effect.

The reason why his sermons seemed inferior to his lectures is very evident. There was no eloquence. There was not one word said for effect. The over-flowing energy and clearness of the lecture-room were replaced by an earnestness of manner, best summed up in the word devoutness. His object seemed to be, to make the most use of the words of Scripture, and to make as little of his own words as he could. Hence a stranger was struck first by the number and rapidity of his references to texts in the Old and New Testaments, and secondly by the devoutness of his manner.

These sermons were always extemporary, but they

were prepared with great care. On the two sides of a card he made the shortest and neatest notes of the texts which he intended to use. One of these cards (see opposite page), taken almost by accident out of very many, will show his manner and his mind better than any description can do.

A friend says: 'I once heard him read the Scriptures at the chapel where he was an elder. He read a long portion of one of the gospels slowly, reverently, and with such an intelligent and sympathising appreciation of the meaning, that I thought I had never heard before so excellent a reader.'

- II. Peter iii. 1, 2, 14. *A prophetic warning to Christians.*

- First the power and grace and promises of the Gospel.
- \. *3.* by His power are given great and precious promises *4* divine
- nature and brethren exhorted to give diligence *5* whilst in this life up
- to *v 8.*

- *Then* cometh a warning of the state into which they *may* fall *8. 9* if
- they forget—so He stirs them up *12. 13. 15.* as escapers from the cor-
- ruption \. *4*

- \\\. *14. Wherefore beloved seeing ye* LOOK *for such things*, their hope
- and expectation—it is to stir up their pure minds \\\. *1.* by way of
- remembrance—hastening the day of the *v 12*   awful as that day will
- be *12.7* because of the deliverance from the plague of our own heart
- II. Cor. iv. 18 17 16 look not at things seen—temporal

- Titus ii. 13 looking for the hope and glorious appearing

  Heb. x. 37 Yet a little while and he that shall come will come.

- The world make His forbearance a plea to forget Him or deny Him
- \\\. *4. 5* perceiving Him not in His works. His people see His mercy
- and long suffering and look for His promise *12 14* and salvation *15*
- and learn that He knoweth how to reserve \\. *3. 9.* and preserve, hence
- they are not to be slothful Prov. xxiv. 30

      nor sleeping—Matt. xxv. 1.   Sleeping virgins

- nor doubting \\\. *4*
- nor repining Heb. xii. 12. 3. 5 lift up hands

        Jas. v. 7. 8 be patient—husbandman waiteth

- but ~~patient~~ waiting Luke 12. 36. 37   39   40   Peter 41

- v. 58 59 refers to day of long suffering

- Wherefore beloved seeing ye know these things, beware &c. danger
- of falling away in many parts \. *9* \\. *20. 21. 22.* great pride of the
- formal adherers \\. *19. 13*

    But the assurance is at \\\. *18—*\. *2. 8*

The figures in italics chiefly refer to the chapter from which the text is
taken.

Figures sloping to the right are for chapters, to the left for verses.

I.

In 1839, very little work was done in the early part of the year in the laboratory. Later, in August, September, October, and November, he worked hard at the source of electricity.

In January 1840 he was again at work on the voltaic pile. He made some experiments in August and September, but after the 14th no entry is made in his note-book until June 1, 1842.

The sixteenth and seventeenth series of 'Experimental Researches' were sent to the Royal Society in the early part of the first year.

Dr. Tyndall sums up this investigation in these words :—

'The memoir on the " Electricity of the Voltaic Pile," published in 1834, appears to have produced but little impression upon the supporters of the contact theory. These indeed were men of too great intellectual weight and insight lightly to take up, or lightly to abandon, a theory. Faraday therefore resumed the attack in two papers communicated to the Royal Society on February 6 and March 19, 1840. In these papers he hampered his antagonists by a crowd of adverse experiments. He hung difficulty after difficulty about the neck of the contact theory, until in its efforts to escape from his assaults it so changed its character as to become a thing totally different from the theory proposed by Volta. The more persistently it was defended, however, the more clearly did it show itself to be a congeries of devices, bearing the stamp of dialectic skill rather than that of natural truth.

'In conclusion, Faraday brought to bear upon it an argument which, had its full weight and purport been understood at the time, would have instantly decided the controversy. "The contact theory," he urged, " assumes that a force which is able to overcome powerful resistance, as for instance that of the conductors, good or bad, through which the current passes, and that again of the electrolytic action where bodies are decomposed by it, *can arise out of nothing*; that without any change in the acting matter, or the consumption of any generating force, a current shall be produced which shall go on for ever against a constant resistance, or only be stopped, as in the voltaic trough, by the ruins which its exertion has heaped up in its own course. This would indeed be *a creation of power*, and is like no other force in nature. We have many processes by which the *form* of the power may be so changed, that an apparent *conversion* of one into the other takes place. So we can change chemical force into the electric current, or the current into chemical force. The beautiful experiments of Seebeck and Peltier show the convertibility of heat and electricity ; and others by Oersted and myself show the convertibility of electricity and magnetism. *But in no case, not even in those of the gymnotus and torpedo, is there a pure creation or a production of power without a corresponding exhaustion of something to supply it.*"'

His work for the Institution in 1839 was eight lectures after Easter on the non-metallic elements, oxygen, chlorine, hydrogen, nitrogen, phosphorus, sulphur, carbon. He gave five Friday evening discourses : On the electric powers of the gymnotus and silurus ; an account of Gurney's oxy-oil lamp, Airy's correction of ships' compasses ; general remarks on flame ; on Hullmandel's calico printing.

In 1840 he gave a course of seven lectures on the force usually called chemical affinity. In his third lecture, speaking of putrefaction and decay, he says : ' Repulsive as these are in some points of view, they are in others full of beauty and of power, and evidences of a wisdom which the more a man knows the more freely will he acknowledge he cannot understand.'

At the end of the fifth lecture he writes : ' There are some strange cases of chemical affinity in which proportions not definite appear to cause great change. Thus silicon and carbon with iron. As there are no exceptions in natural laws, we shall probably hereafter find these important. Observe a case therefore in steel. Steel is soft iron combined with a little carbon. Now observe the similarities and differences.' And a dozen experiments are given in illustration of these. ' Reference of this case to those of organic life, i.e. the effect of a small portion of matter—so make us modest and doubtful in our assertions. Privilege of the experimentalist. The need of this modesty shown still more by the present state of the notions regarding atoms, and grouped atoms, and multiple atoms, &c.' He also gave three Friday evening discourses on voltaic precipitation, electrotype ; on condensed gases ; and on the origin of voltaic electricity. The following note shows how at this time he arranged the Friday evening meetings and prepared his own lecture.

FARADAY TO C. WHEATSTONE, ESQ.

'Royal Institution : May 4, 1840.

' Dear Wheatstone,—Thrice have I endeavoured to catch you at home, but failed. My object is to ascertain whether you can let me have the telegraph subject this season, as you said at the beginning of the year ; and if so, whether the last evening, June 12, would suit. As I have not seen much of it lately *I should want cramming*, but will prepare as soon as I know your mind. You know my desire to present your beautiful developments to our audience, and I know I may count on your willingness, but cannot tell as to your convenience. Let me know quickly, for I must now arrange the rest of the evenings. Ever yours,

' M. FARADAY.'

His work for the Trinity House was very little in 1839. At the end of July he was four days at Orfordness, measuring and comparing at sea and on land the Argand lamp, the French lamp, and the Bude lamp.

In 1840 he reported to the Trinity House on the necessity and method of examining lighthouse dioptric arrangements, and he had to examine the apparatus intended for Gibraltar. Between Purfleet and Blackwall he made a long comparison between English and French reflecting lamps, and between English and French refracting prisms.

II.

In 1840, he was made Member of the American Philosophical Society, Philadelphia ; and Honorary

1839–40. Member of the Hunterian Medical Society, Edin-
Æt.47–49. burgh.

III.

He showed something of his nature in his letters
to Dr. Hare, Professor of Chemistry in the University of Pennsylvania, and to Professor Auguste de la
Rive, the son of his early friend. In 1840 Dr. Hare
wrote his objections to Faraday's theoretical opinions on
static induction. At the end of Faraday's reply, he
says:—' The paragraphs which remain unanswered
refer, I think, only to differences of opinion, or else
not even to differences, but opinions regarding which I
have not ventured to judge. These opinions I esteem
of the utmost importance, but that is a reason which
makes me the rather desirous to decline entering upon
their consideration, inasmuch as on many of their connected points I have formed no decided notion, but
am constrained by ignorance and the contrast of facts to
hold my judgment as yet in suspense. It is indeed to
me an annoying matter to find how many subjects there
are in electrical science on which, if I were asked for
an opinion, I should have to say I cannot tell—I do
not know; but, on the other hand, it is encouraging to
think that these are they which, if pursued industriously,
experimentally, and thoughtfully, will lead to new discoveries. Such a subject, for instance, occurs in the
currents produced by dynamic induction, which you
say it will be admitted do not require for their production intervening ponderable atoms. For my own
part, I more than half incline to think they do require these intervening particles. But on this question,
as on many others, I have not yet made up my mind.'

On January 1, the following year, Dr. Hare sent a reply. In Faraday's answer to this, he says :—' You must excuse me, however, for several reasons, from answering it at any length. The first is my distaste for controversy, which is so great that I would on no account our correspondence should acquire that character. I have often seen it do great harm, and yet remember few cases in natural knowledge where it has helped much either to pull down error or advance truth. Criticism, on the other hand, is of much value ; and when criticism such as yours has done its duty, then it is for other minds than those either of the author or critic to decide upon and acknowledge the right.'

1839–40.
Æt.47–49.

FARADAY TO PROFESSOR AUGUSTE DE LA RIVE.

'Royal Institution : April 24, 1840.

' My dear Sir,—Though a miserable correspondent, I take up my pen to write to you, the moving feeling being a desire to congratulate you on your discernment, perseverance, faithfulness, and success in the cause of *Chemical Excitement of the Current in the Voltaic Battery.* You will think it is rather late to do so ; but not under the circumstances. For a long time I had not made up my mind ; then the facts of definite electro-chemical action made me take part with the supporters of the chemical theory, and since then Marianini's paper with reference to myself has made me read and experiment more generally on the point in question. In the reading, I was struck to see how soon, clearly, and constantly you had and have supported that theory, and think your proofs and reasons most excellent and convincing. The constancy of Marianini and of many others on the opposite side made me, however, think it not unnecessary to accumulate and record evidence of the truth, and I have therefore written two papers, which I shall send you when printed, in which I enter under your banners as regards the origin of electricity or of the current in the tube. My object in experimenting was, as I am sure yours has always been, not so much to support a given theory as to learn the natural truth. And having gone to the question unbiassed by any prejudices, I cannot imagine how anyone whose mind is not preoccupied by a theory, or a strong bearing to a theory, can take part with that of contact against that of chemical action. However, I am perhaps wrong saying so much, for, as no one is infallible, and as the experience of past times may teach us to doubt a theory which seems to be most unchangeably established, so we cannot say what the future may bring forth in regard to these views.

' I shall be anxious some day, if health continues, to make a few experiments on contact with the electrometer. I know of yours, Becquerel's, &c., but if there are any dimensions which are particular, or any precautions which as a practical man you are aware of, and which you know render it more sensible, I am in hopes, if you take the trouble to write to me hereafter, you would not mind sending me word or referring me to the papers or works which may mention them.

' And now, before I conclude, let me ask you to remember me kindly to Madame de la Rive, of whose good will and courtesy both I and my wife have a very strong remembrance. I was not well during my journey at that time, but still I have a great many pleasant recollections, and amongst the most pleasant those of Geneva, for which I am indebted to you.

'I have several papers of yours to acknowledge, but I cannot recollect them so accurately as to thank you in order for them. I am always grateful, and very glad to see them. Your historical account of your own researches as regards the battery has been very useful to me, and makes me wish more and more that we had a sort of index to electrical science to which one might look for facts, their authors, and public dates. The man who would devise a good scheme for such an index, so that it might take in new facts as they were discovered, and also receive old and anticipating observations as they should gradually be remembered and drawn forth from obscurity, would deserve well of all scientific men.

'But I must conclude, and am as ever, my dear Sir, your obliged, grateful, and admiring friend,

'M. FARADAY.'

The following letter from Professor Brande to Faraday shows how his work was telling on his health, and it gives his relationship at this time to Mr. Brande.

'Royal Mint: December 11, 1839.

'My dear Faraday,—Many thanks for your very satisfactory note, which I assure you has given all of us sincere pleasure. I called yesterday morning on Dr. Latham to tell him its contents, and ask whether he had any further advice to give you. He says, none, provided you will continue as you have begun, and remain thoroughly idle. Pray act strictly upon this principle. You can have no difficulty in amusing and occupying yourself with what you call trifles, things which do not require thought or consideration ; your brain will then regain its tone, and you will be able to make moderate and prudent use of its faculties. Dr. Latham expressed his sincere conviction that under these conditions all would get quite right again. It grieves me that I cannot offer to be of any use to you as regards the Friday evenings, but you know how sad a figure I cut on those occasions ; and as to the tact requisite for their general management and arrangement, I candidly confess I have it not. However, I will do all and anything I can, and if you will suggest anything which I can follow up, or point out any inquiries which I can make, or persons to whom I can apply, you have only to send me your hints and orders. Not but that I confidently hope that before their time comes on you will feel quite up to all business of that kind. I got on tolerably well with the electricity lectures. In Anderson I have an excellent prompter ; he tells me that I do better than he would have expected—a plain compliment which I duly appreciate. At first I began to fear the fate of Phaeton in the chariot of Phœbus, but by now and then going a little astray from your notes, and following the excellent maxim of not attempting, as the metaphysicians do, to explain what I do not understand, I hope I shall not commit myself. I admire your apprehension of having ridden your hobby in improper times and places ; no one could say you were not his master. I am afraid they sometimes see that I am mounted upon an animal I am afraid of. You would have been amused the other day had you been present at the Athenæum House Committee upon the subject of illumination. The old gas apparatus for that purpose is worn out, and it has become necessary to replace it ; the question therefore naturally arose, as to whether the arrangement and device might not be improved, and it was thought right to consult an artist or two: accordingly, those two excellent persons, as well as artists, Sir A. Callcot and Sir F. Chantry, were applied to; the former suggested placing Minerva in a niche of lights, and the latter adopted the notion as a good one and gave a plan for the purpose ; it was thought original and highly appropriate. But when we (the common plain members of the committee) came to examine the matter, we found that Minerva would probably have been redhot before the evening was over, or more likely blistered, splintered, or fused ; for, to add to the joke, we found on inquiry, that, to render her *waterproof*, she had been imbued with wax. Now although Minerva's power was very great, and her attributes superhuman, I do not remember among them that of being *fireproof*. I shall have some fun with Sir Francis upon this matter.

'I have not said half that I had intended, but have filled my paper. Make my kind regards to Mrs. Faraday, take care of yourself, and when you feel inclined oblige me with a line or two, and set me about anything I can do for you to relieve you of trouble.

'Yours, dear Faraday, very sincerely,

'W. T. B.

'Professor Faraday, 80 King's Road, Brighton.'

Faraday himself drew up the following table of the work he had given up during the first period of his experimental researches in electricity. This period lasted for ten years.

1840.
Æт.48-49.

	Gave up Friday Evenings.	Gave up Juvenile Lectures.	Gave up Mr. Brande's 12 morning lectures.	Closed three days in the week. (Saw no one.)	Declined reprinting 'Chemical Manipulation.'	Gave up many morning lectures.	Gave up the rest of Professional business.	Gave up the excise business.	Declined all dining out or invitations.	Gave up Professional business in Courts.	Declined Council business of the Royal Society.
Æt. 42–3, 1834.											
1834.											
Æt. 43–4, 1835.											
Æt. 44–5, 1836.											
Æt. 46, Nov. 1837.											
Æt. 46–7, 1838.											
1838.											
Æt. 48, Nov. 1839.											
Dec. 1839.											
Æt. 49, Dec. 1840.											
Æt. 50, Jan. 1841.		May gave up Easter lectures and all other business at R.I.									

His niece, Miss Reid, who lived with Mr. and Mrs.
Faraday at the Institution, has thus given her recollections of him from 1830 to 1840:—

'There could be very few regular lessons at the
Institution, there were so many breaks and interrup-

tions. Sometimes my uncle would give me a few sums to do, and he always tried to make me understand the why and wherefore of everything I did. Then occasionally he gave me a reading lesson. How patient he was, and how often he went over and over the same passage when I was unusually dense! He had himself taken lessons from Smart, and he used to practise reading with exaggerated emphasis occasionally.

' In the earlier days of the Juvenile Lectures he used to encourage me to tell him everything that struck me, and where my difficulties lay when I did not understand him fully. In the next lecture he would enlarge on those especial points, and he would tell me my remarks had helped him to make things clear to the young ones. He never mortified me by wondering at my ignorance, never seemed to think how stupid I was. I might begin at the very beginning again and again; his patience and kindness were unfailing.

' A visit to the laboratory used to be a treat when the busy time of the day was over.

' We often found him hard at work on experiments connected with his researches, his apron full of holes. If very busy he would merely give a nod, and aunt would sit down quietly with me in the distance, till presently he would make a note on his slate and turn round to us for a talk; or perhaps he would agree to come upstairs to finish the evening with a game at bagatelle, stipulating for half an hour's quiet work first to finish his experiment. He was fond of all ingenious games, and he always excelled in them. For a time he took up the Chinese puzzle, and, after making all the figures in the book, he set to work and produced a new set of figures of his own, neatly drawn, and perfectly accurate in their proportions, which those in the book were not. Another time, when he had been unwell, he amused himself with *Papyroplastics*, and with his dexterous fingers made a chest of drawers and pigeon-house, &c.

'When dull and dispirited, as sometimes he was to an extreme degree, my aunt used to carry him off to Brighton, or somewhere, for a few days, and they generally came back refreshed and invigorated. Once they had very wet weather in some out-of-the-way place, and there was a want of amusement, so he ruled a sheet of paper and made a neat draught-board, on which they played games with pink and white lozenges for draughts. But my aunt used to give up almost all the games in turn, as he soon became the better player, and, as she said, there was no fun in being always beaten. At bagatelle, however, she kept the supremacy, and it was long a favourite, on account of its being a cheerful game requiring a little moving about.

'Often of an evening they would go to the Zoological Gardens and find interest in all the animals, especially the new arrivals, though he was always much diverted by the tricks of the monkeys. We have seen him laugh till the tears ran down his cheeks as he watched them. He never missed seeing the wonderful sights of the day—acrobats and tumblers, giants and dwarfs; even Punch and Judy was an unfailing source of delight, whether he looked at the performance or at the admiring gaping crowd.

'He was very sensitive to smells; he thoroughly enjoyed a cabbage rose, and his friends knew that one was sure to be a welcome gift. Pure eau-de-Cologne he liked very much; it was one of the few luxuries of

the kind that he indulged in ; musk was his abhorrence,
and the use of that scent by his acquaintance annoyed
him even more than the smell of tobacco, which was
sufficiently disagreeable to him. The fumes from a
candle or oil-lamp going out would make him very
angry. On returning home one evening, he found his
rooms full of the odious smell from an expiring lamp ;
he rushed to the window, flung it up hastily, and
brought down a whole row of hyacinth bulbs and
flowers and glasses.

'Mr. Magrath used to come regularly to the morning
lectures, for the sole purpose of noting down for him
any faults of delivery or defective pronunciation that
could be detected. The list was always received with
thanks; although his corrections were not uniformly
adopted, he was encouraged to continue his remarks
with perfect freedom. In early days he always lectured
with a card before him with *Slow* written upon it in
distinct characters. Sometimes he would overlook it
and become too rapid ; in this case Anderson had
orders to place the card before him. Sometimes he
had the word "Time" on a card brought forward
when the hour was nearly expired.'

APPENDIX I.   CHAPTER I.

— ◦◊◦ —

The following letters are necessary to complete the story of the pension; they are full of traits of Faraday's character.

FARADAY TO SIR JAMES SOUTH.

'Royal Institution: Nov. 6, 1835.

'My dear Sir James,—Need I say how thankful I am to you for your letter and Miss Fox's approval?   Now I have no anxiety: those whose kindness I am grateful for, and whose approbation I am anxious to have, are with me, and I hope that in doing what was right I have not given others occasion to have one evil thought of me.

'Since I first knew of the affair, nothing has been nearer to my mind than the desire, whilst I preserved my self-respect, to give no one occasion of offence.

'As you have been the bearer of Miss Fox's kind expressions to me, will you do me a last favour, by placing in her hands a few words of thanks?   I think I ought to send them, only hoping that in this and other things I have not been too much obtruded on her attention.

'And now, my dear Sir, pray *let me drop*.   I know you have serious troubles of your own.   Do not let me be one any longer, either to you or to others.   You have my most grateful feelings for all the kindness you have shown to him who is ever truly yours,

'M. FARADAY.'

TO THE HONOURABLE MISS FOX.

'Royal Institution: Nov. 6, 1835.

' Madam,—My feelings prompt me, and the sight of your
handwriting encourages me, to offer you in few words my
most grateful and sincere thanks for the extraordinary kindness which you have shown to me in a late affair, the conclusion of which is to me a source of pleasure, since it allows me
to express my feelings without any fear of a mistaken interpretation being put upon them.

' Your kind expressions and invitation I do not merit, and
it is very probable that, upon a nearer view of me, you might
think that your present estimate of my character is much too
favourable. But I shall never forget that what you know of
me thus far has gained your approbation, and it will be doubly
my desire and pleasure henceforward to deserve and retain it.

' I have the honour to be, Madam, with the most sincere
respect, your grateful and humble servant,

' M. FARADAY.'

FARADAY TO THE LADY MARY FOX.

'Royal Institution: Nov. 23, 1835.

' My Lady,—Permit a very humble person to intrude for
a moment on your Ladyship's attention, that he may return
his heartfelt thanks for all the kindness you have shown him.
I scarcely know what terms to use by which I may express
the sincerity of those thanks ; but I will trust to the same
kind heart which was ready to think well of one who was not
known to you in person or in act, for a favourable interpretation of my feelings at the present moment.

' Deep and anxious thoughts have delayed my acknowledgment of your Ladyship's kindness, which, however, in no
state of mind can I ever forget, and which I trust I shall ever
strive to deserve.

'I have the honour to be, my Lady, your Ladyship's most grateful and humble servant,

'M. FARADAY.'

DR. HOLLAND TO FARADAY.

'Lower Brook Street: Nov. 29, 1835.

'My dear Sir,—I saw Lord Melbourne this morning. He begged me particularly to mention to you that the King had expressed great satisfaction in the arrangement made as to the pension. This he was sure it would be gratifying to you to learn.

'We spoke of the paper which has appeared in the "Times;" with the same generous frankness which I have found in him throughout, he begs you not to be disgusted by this in reference to him. Though regretting the circumstance, he was prepared for the likelihood of these things being converted to party purposes. He concurs in the impression I had myself ventured to form ; that it is better (at present at least) to trust for refutation to the simple and conscientious declarations which you, I, and others, are able to make on the subject, whensoever inquiries or comments give opportunity for this. I would fain hope that even yet truth and justice may be satisfied by the affirmation we can explicitly make, that honour and generous feeling have been preserved throughout.

'Believe me, my dear Faraday, yours very faithfully,

'H. HOLLAND.'

FARADAY TO THE RIGHT HON. LORD HOLLAND.

'Royal Institution : December 1, 1835.

'My Lord,—I have seen Dr. Holland, who suggests to me to write to your Lordship, and say that he thinks no steps of a public nature ought to be entered upon without Lord Melbourne's consent, since it might lead to discussions of which we cannot see the end. If Lord Melbourne should be of a different opinion, then, on communicating with Dr. Holland, he

with Sir James South will be able to take such steps as may   1835.
under existing circumstances appear to be most expedient.   Æt. 44.
' I have the honour to be, my Lord, your Lordship's most
obedient, humble servant,

'M. FARADAY.'

DR. HOLLAND TO FARADAY.

Lower Brook Street : December 2, 1835.

'My dear Faraday,—I very reluctantly write anything
further, on a subject which I know to be harassing to you;
but I cannot do otherwise than forward this letter of Lord
Holland's, which I have just received. It is manifest that he
considers (as Miss Fox and some others have done) that Lord
Melbourne's indifference to his own public vindication from
this injurious paper is not to be taken as a justification for
omitting this ; and possibly they are right, though I still
incline to think that such explanation of the truth might have
been attained by private statement, made fully and explicitly
whenever occasion occurred.

' But as there is doubt on the subject, might not a very
brief letter be written under some such form as Lord H.
suggests, or beginning thus, " That a published statement had
been put into your hands, very inaccurate in many respects,
and likely to convey impressions very different from the truth,
that though not entitled to state the details of what had passed
between Lord M. and yourself, in the progress of this transaction, yet you felt it right, &c. ; " and then to state briefly
your sense of Lord M.'s honourable and generous conduct,
and of the satisfaction it gave to your own feelings in the
transaction? I do but simply suggest, however, what, if you
judge it right to do this, you will much better express in your
own words ; there would be this advantage in doing it, that
I think it would assuredly and completely *close* the whole
business. Everything privately said by us would be in full
concurrence with it, and at the same time with still more
explicit declaration of the high sense of honour shown by
yourself in every part of the transaction. Nor do I think

another word could be said elsewhere on the subject. I state everything as it occurs to me, and would rather have sought opportunity of doing it by calling upon you, had I not been engaged with patients in other directions this evening. You must excuse, therefore, a hastily written note, and believe me, ever yours, my dear Faraday,

'H. HOLLAND.

' I presume Lord H. will see Lord Melbourne to-day or to-morrow, but he speaks as if he perceived with certainty that Lord M. could not object to what he himself suggests.

'I trust you will not hesitate from fear of troubling me, if you wish that we should meet for further communication on the subject, or to arrange a meeting with Sir J. South.'

FARADAY TO DR. HOLLAND.

'Royal Institution : December 2, 1835.

'My dear Sir,—Your letter distresses me, for I thought that mine to Lord Holland would have intimated that I was willing to do all I ought to do to relieve others from inconvenience (though in no way the cause of it), but that I durst not in doing so place myself in a false and dangerous position. This I should be doing were I to proceed without Sir James South. The affair is actually more his than mine; he is more concerned in the matter, and he *knows more about it* than I do; his character is as much at stake as mine; he has been accepted by Lord Melbourne through you as my representative. I am willing to do anything that he and you may advise, for I feel sure that neither would propose what I could not assent to. Let me then say distinctly, that I cannot take a single step in this affair without him, and that it is with him, and not with me, that Lord Melbourne's friends should communicate. Let me pray you, therefore, to communicate Lord Holland's letter (which I return) to him, and I shall wait until you two tell me what to do.

'Excuse me, my dear Sir, for speaking plainly in this matter, but I perceive that unless I do so, very serious consequences may arise. The pension is a matter of indifference to me, but other results, some of which have already come to pass, are not so; the continued renewal of this affair to my mind, and that in a manner hesitating, dilatory, and changeable, is not consistent with my feelings, and tempts me at times to what might be thought very ungenerous under the circumstances, namely, even at this late hour, a determined refusal of the whole.

1835.
ÆT. 44.

'Trusting, however, that all these difficulties will vanish upon your communicating with Sir James South,

'I am, my dear Sir, your truly grateful and obliged

'M. FARADAY.'

FARADAY TO MISS FOX.

'Royal Institution: December 12, 1835.

'My dear Madam,—Accept my most grateful thanks for your kind letter which I have just received. Most sincerely do I regret that I cannot comply with the wish expressed in it, as I am persuaded that any statement of mine in the public papers cannot NOW be attended with any advantage to either party.

'Repeating my heartfelt gratitude to you for the kind interest you have shown me during the whole transaction, I have the honour to subscribe myself, my dear Madam, your most obedient, faithful servant,

'M. FARADAY.'

This letter is in Sir James South's writing.

FARADAY TO THE RIGHT HON. LORD VISCOUNT MELBOURNE,
FIRST LORD OF THE TREASURY.

'Royal Institution: December 30, 1835.

'My Lord,—I intrude, perhaps improperly, to thank your Lordship again for the great kindness and condescension which your Lordship has shown me in the late affair of the pension, and to ask a further grace; the good opinion of me which your Lordship has expressed as being the foundation of the former affair, encouraging me to hope I shall not be considered intrusive in the present.

'Professor Magnus of Berlin has sent me some potassium, sodium, and other articles in a tin box enclosed in a package consigned to Messrs. Hamilton, Koch, and Co. The tin box, the contents of which are entirely for purpose of philosophical investigation, has been stopped at the Custom House, and the favour I have to ask of your Lordship is that it may be allowed to pass—the things that are not in the tin box pay duty regularly. The contents of the box, if I had to pay for them at Berlin, would not cost me, I believe, more than 3*l*. or 4*l*., but it is their peculiarity which makes the difficulty at the Custom House.

'I have the honour to be, my Lord, your Lordship's most obliged and grateful servant,

'M. FARADAY.'

LORD MELBOURNE TO FARADAY.

'Panshanger: January 2, 1836.

'Sir,—I have again to express my gratification that the matter of the pension terminated in so amicable a manner, and I beg to return you my thanks for your willingness, expressed through Dr. Holland, to contradict any injurious statements in the public prints. The best course was followed, which was to terminate the discussion as soon as possible. I will give directions to have everything that is possible done at the customs about your box from Berlin, and as speedily as possible.

'Believe me, yours faithfully,

'MELBOURNE.'

FARADAY TO MISS FOX.

'Royal Institution: January 30, 1836.

'Dear Madam,—You will be glad to hear that Lord Melbourne, who lately did me the great favour to pass some foreign scientific articles at the Custom House which would not have been allowed but for such authority, in a letter with which his Lordship honoured me, expressed his approbation of the course pursued in a late occasion as being that which was decidedly the best. This was the spontaneous opinion of his Lordship, there being no allusion to that course on my part.

'With the sincerest wishes for your health and happiness, I am, dear Madam, your much obliged and faithful servant,

'M. Faraday.'

MISS FOX TO FARADAY.

'Ordnance Office : Tuesday evening, February 16, 1836.

'Dear Sir,—On returning home, after an absence of more than two months, I found your letter. . . I am very glad indeed, and not at all surprised, to hear of Lord Melbourne's civility ; he is no egotist—frank, generous, and manly ; and you must allow me to say that in a case where an unfair impression has been given of the conduct of a man of such a character, he is the very last person to be consulted, or to require the contradiction of an aspersion which he is conscious is utterly undeserved; but his friends probably did, and do still, think otherwise, though now certainly the time is gone by for drawing public attention to the subject.

'I am, dear Sir, with very true respect and regard, yours,

'Caroline Fox.'

APPENDIX II.

1. *
2. * Water on top of tube heated—then connected with upper vessel and currents
3. * Flask of water with indigo below heated   5'

4. * Copper kettle over ring gas flame—for simmering
5. * Water under air-pump to evolve air
6. *
7. * Steam from our boiler through a pipe in a jar of water
8. *   12'

9. * Hot thick capsule—spheroidal water—thermometer in 208° F.
10. * Not touch. Iodine and its effects
11. * Hot piece of silver in little water
12. * Potassium in water residue and final explosion   20'

13. * Bent tube with water — standing of the column — water hammer
14. * Water crackling in oil flame, large, tow on wire
15. * Wet paper into oil heated in a tube
16. *
17. *   30

18. *
19. * Warm water under air-pump
20. * Water in retort boiling—cork up and dip in cold water   40'

21. * Chinese or Indian saucepans. Savages, Newfoundlanders, drawing
22. * Pipkins—enamelled vessels
23. * Boil water in paper by steam—cook chestnuts in box 15 or 20   45'

24. * Furred kettle—deposits—eggs, &c.   San Fillippo casts
25. * Distilled water and fur in flask boiling—add muriate of ammonia and test
26. * Two flasks boiling—add muriate of ammonia to one—clear it   55'

27. * Flask of water and gold leaf—bucket to collect
28. * Model of Perkins's boiler   60'

29. * Large clean flask of acid water—filings in

30. * Gold paper or leaf on finger and hot ball—black pigment
31. *
32. *

CHAPTER I.

Fine extension of current principle 2   all parts heated 3

Effects of the heat—*Simmering* 4   Tea-urn, dependent upon the
air naturally in the water 5
      Boiling—the *fixed temperature*—in cooling, &c.
The heat *lost or stored* in the steam 7

Consider the point of change   in open air or pressure of air
First, is lowest in the hottest vessels. Boutigny, 205° F.
      9   10   11   12
Next, ordinary state, 212°. Varying a little with vessels.

Next pure water and its temperature. Denny, 275° F. 3 atmo-
      spheres   13   14   15
Wonderful to see how the true point in water is never the ordinary
point—and how we are protected by comparatively trifling circumstances.   Besides these considerations

---

Effect of pressure on the boiling point
Increase
Diminution   19   20

Substance of boilers and modes of boiling
      21   22   23

Fur deposits 24   and clearance 25   26

Places of rest in boilers 27   and their use 28

Nuclei 29

Black bottoms to vessels 30

Exception to expansion in water.

CHAPTER II.

1841.

Æt.49-50.

Loss of memory and giddiness had long, occasionally, troubled Faraday, and obliged him to stop his work. But now they entirely put an end to all his experiments. For four years, with the exception of an inquiry into the cause of the electricity produced by a jet of steam, no experimental researches in electricity were made. For a year he rested almost entirely, he gave no lectures, and he went for three months to Switzerland. After a year he began again to work for the Institution, and when he did go on with his researches, he returned to the liquefaction of the gases.

In different ways he showed much of his character during this period of rest. The journal he kept of his Swiss tour is full of kindness, and gentleness, and beauty. It shows his excessive neatness. It has the different mountain flowers which he gathered in his walks fixed in it, as few but Faraday himself could have fixed them. His letters are free from the slightest sign of mental disease. His only illness was overwork, and his only remedy was rest.

Almost the only work he did in 1841 was for the Trinity House. For the Royal Institution he did all he could ; he gave the Juvenile Lectures at Christmas on chemistry.

On February 2, he went down to St. Catherine's lighthouse in the Isle of Wight, to remedy the condensation of moisture on the glass in the inside. On the 6th he returned home, 'quite satisfied with the chimney, and have no doubt we shall have a lantern quite clear from sweat, and also much cleaner, both as to the mirrors and roof, from soot and blackness, than heretofore.'

This was the beginning of his system of ventilation in lighthouses, and gradually he had it carried out in all the lighthouses of England. He did other work for the Trinity House in May and June, but nothing of importance.

The only honour he received was an offer to lecture in America.

Mr. J. A. Lowell, of Boston, U.S., invited him to give twelve lectures at the Boston Institution, on chemistry and its applications. ' I should not have presumed to apply to Mr. Faraday,' he said, 'had not Dr. Warren reported that Mr. F. had expressed a wish to visit this country. The remuneration would be 450*l.*'

The Swiss Journal begins thus:—

Wednesday, June 30th.—Left London at half-past six o'clock by steamboat. Had fine day to Ostend. Arrived there about half-past six P.M. (Mr. George Barnard and Mrs. Barnard and Mr. and Mrs. Faraday.)

Sunday, July 4th.—Stopped at Aix; comfortable in our apartments; a busy feast; procession day in the town in honour of St. Peter, or rather of a church called after his name; many priests, women and children dressed in white, men, &c., with much singing and music: the procession lasted many hours; flowers were strewn in the way, garlands hung across the streets, whole trees cut down, and stuck up at the doors, &c.; a quiet walk to Borcette.

Monday, 5th.—Left Aix in a carriage, and reached Cologne by seven o'clock P.M. . . . . Bought some eau-de-Cologne at one of the thirty veritable manufacturers; took some trouble with George to find out the shop, which was the wrong one; the right one was so evident, we thought it suspicious; inn very noisy all night, but the view from our window beautiful—moon and river.

Tuesday, 6th.—Took steamboat at seven o'clock this morning, and are now at Coblentz, having done this by four o'clock. . . . . Ehrenbreitstein is just now gloriously illuminated by the setting sun, glowing as if on fire. Have walked out to see the Moselle and bridge over it; also that fine specimen of honour and glory set up jointly by the French and Russian authorities in succession. It is an excellent illustration of the word that all is vanity and vexation of spirit. Rains now.

Wednesday, 7th.—Left Coblentz by steamboat at six o'clock, and passed the most beautiful part of the Rhine. Is certainly wonderful, and the castles! what scenes must have been witnessed there in former times!

Thursday, 8th.—The morning found us still on the water, for going against the stream is hard work; the river very diffuse; many islands and overflowings; country flat and uninteresting, but rich and verdant; at about half-past four reached Strasburg, and put up at a good inn; went to see the wonderful cathedral; ascended the tower, saw the market-place, the fine old houses, the storks on their summits: it being evening time, they came in, alighting on the sharp high roofs as I looked at them. The cathedral was in the finest state possible, and the houses in perfect harmony with it; glorious clouds and lights; good weather to-day, but some showers.

Friday, 9th.—Ran about the town a little, in fact all round it; the storks very interesting, rising from their nests, and the markets all in the highest activity. At the fish market, all the fish were alive, and swimming in flat vessels containing water. When sold, they were weighed in nets, and then killed by a blow. Prepared frogs were very abundant, the hind legs being strung on small willows and sold in bundles, or rather rows, and the bodies in a separate parcel; of the latter, however, there were not so many. Went into the cathedral; I like it better within by evening light; I think the glass looks better; when the sun is full on a window it is not so good as when the shadow of a buttress falls on parts; and as the lights decay towards evening, they sink faster on opaque objects than on transparent ones, seen from within, and thus the contrast between the walls and the windows is much greater in the evening; but the windows are exceedingly beautiful at any time. Service was going on in various parts, and much food for reflection supplied by the people. The inside of the cathedral disappoints me; the outside is wonderful. I think it could not have been constructed with any other stone than the fine-grained sandstone which has been employed. It unites that strength, durability, and facility of working which were essential for a work at once so large in its mass, and so delicate in its details. As a whole, the exterior of the cathedral disappoints me, i.e. its general form is meagre and poor in effect, but its parts are wonderful; its height about 500 feet, and so perpendicular.

Saturday, 10*th*.—Left Bâle by voiturier, on our way to Berne. The day rainy throughout, but on the whole this not bad for us. For as our way lay through the valley of the Münsterthal, the principal scenery was close scenery, and we had water effect in perfection. The river by the road was full, and running over in places; trees and a door I saw sweeping down on its surface; the waters very turbid, streams and jets of water on all sides of us; fine cascades; perpendicular or overhanging rocks of the most beautiful characters, some 1,200 or more feet high, and the trees growing out of the crevices in a most curious way; in one place, the perpendicular surface of a vast rock was so moistened by the water running over it from the land above, and so making some parts darker than others, as to give the appearance of churches, houses, and buildings, and at first I thought a town was before me in the distance; the strata were in many places vertical, and the rocky ribs most wonderful. We arrived at Moutiers in good time, and purpose stopping here to-morrow.

Sunday, 11*th*.—Rested in this pretty Swiss village, which is in the canton of Berne. At half-past five woke up by the horn of the cow-herd, gathering his charge together, as they issued from the doors of the different houses, each with a bell, and conducted them one way out of the town. In a short time afterwards, the shepherd moved by, and the sheep appeared at the doors and in the road; the goats came forth with them, at the call of a goat-herd, who tooted a cow's 1841. horn; these mingled together at the corner place by Æt. 49. the inn until the two men moved towards different directions, when the sheep and goats spontaneously separated, winding in and out amongst one another, and the goats straggled away up the hill, followed by their keeper, whilst the sheep followed their guide into the valley below; it reminded one strongly of the parable.

We passed the day very quietly here; much rain, but still appearing to clear and mend; about four o'clock, the cattle and sheep and goats came home again, and it was pleasant to see them and their owners meet again, the creatures anxious for their homes and the kind care of their owners. The people seem to treat their horses, cattle, &c. kindly, and the animals, consequently, seem to have more intelligence and more freedom of action, which they show distinctly, for if anything occurs, they look on as if with considerable interest, to examine and observe the consequences.

Tuesday, 13*th*.—About two o'clock started (from Berne) for Thun, and after three or four hours' drive arrived here, having had both rain and fine weather on the road. These Alps grow in beauty greatly as one approaches them, and we have had the finest mixtures of these and the clouds which man can imagine, the latter at last dissolving in rain and rejoining the earth. The beauty increased as we advanced up, not only to Thun, but to the Hotel Belle Vue, to which we came. Here we rest for a week, or two or three perhaps, and are delightfully situated, having the upper floor of a house separate from the hotel, and opposite the lake and mountains. We may be as quiet as mice here, if we like; George is in high glee with the tones of the scenery, and means to make much of it. Our dear wives also enjoy the thought of the week's rest and pleasure.

Wednesday, 14*th.*—Having come to a home, our hours are to be very regular, for irregularity is uneconomical; we breakfast at seven in our house, dine at one o'clock at the *table d'hôte* in another house, tea at seven, or thereabout, and propose going to bed at nine o'clock; as for supper, we don't think of it; we do not find the many rolls and loaves too much. As for the fashion of the dinner, it is anything but agreeable, to me, but it satisfies the appetite, and that is a great point.

The river and lake are beautiful to-day, and the mountains also. George made a regular artistical examination of the town and neighbourhood to-day, and I went with him, imbibing the picturesque; there is certainly plenty of it; the morning was sunny and beautiful, and the afternoon was stormy, and equally beautiful; so beautiful I never saw the like. A storm came on, and the deep darkness of one part of the mountains, the bright sunshine of another part, the emerald lights of the distant forests and glades under the edge of the cloud were magnificent. Then came on lightning, and the Alp thunder rolling beautifully; and to finish all, a flash struck the church, which is a little way from us, and set it on fire, but no serious harm resulted, as it was soon put out. This evening it still continues raining, but the lightning is over.

Friday 16*th.*—Took a long walk to the valley called the Simmenthal, which goes off from the valley

of the lake; the day was fine, and I made about
twenty-five miles there and back.

.   .   .   .   .   .

The frogs were very beautiful, lively, vocal, and intelligent, and not at all fearful. The butterflies, too, became familiar friends with me, as I sat under the trees on the river's bank. It is wonderful how much intelligence all these animals show when they are treated kindly and quietly; when, in fact, they are treated as having their right and part in creation, instead of being frightened, oppressed, and destroyed.

Monday, 19th.—Very fine day; walk with dear Sarah on the lake side to Oberhofen, through the beautiful vineyards; very busy were the women and men in trimming the vines, stripping off leaves and tendrils from fruit-bearing branches. The churchyard was beautiful, and the simplicity of the little remembrance posts set upon the graves very pleasant. One who had been too poor to put up an engraved brass plate, or even a painted board, had written with ink on paper the birth and death of the being whose remains were below, and this had been fastened to a board, and mounted on the top of a stick at the head of the grave, the paper being protected by a little edge and roof. Such was the simple remembrance, but nature had added her pathos, for under the shelter by the writing, a caterpillar had fastened itself, and passed into its deathlike state of chrysalis, and having ultimately assumed its final state, it had winged its way from the spot, and had left the corpse-like relics behind. How old and how beautiful is this figure of the resurrection! surely it can never appear before our eyes without touching the thoughts.

Wednesday, 21*st*.—Fine morning, roused early by the village (Fruchtigen) goat-herd, blowing his shell (for it was a shell he used for the purpose), and the goats came bleating out of the houses. The flock which he gathered was a good-sized one. In the night I was roused by the watchman singing out the hour and certain lines of salutation, very pleasant as expressing kindly feeling in a kind tone, but it was perhaps not necessary to waken one out of sleep for such a purpose. After breakfast we walked off, and in two hours and a half, mounting most of the distance, we came to Kandersteg, nearly at the top of the valley of the Kander.

After refreshment here we walked up a valley rising very rapidly to the small lake Oeschinen, which we found enclosed by mountains, many of them covered with snow, and into which were falling eight streams of water in different places from the snow above. Here was very grand scenery. Rocks high and precipitous. Torrent courses. The Alpine rose or rhododendron, very beautiful. Monkshood also very fine growing in the water, and in many places the whole plant, though in full flower, was entirely under the water, standing upright, and looking very healthy and proper. While ascending to this place I was very tired, but a little rest restored me. The call which rising or climbing makes upon the lungs is very distinct and remarkable, and the air here is so attenuated that the same amount of breathing does not do half the good it would below. George sat down to sketch. I rambled about awhile. The courses of the torrents or avalanches here are marked by very striking appearances ; a long line of pines swept down and broken or splintered in every possible way, but all in one direction, give sufficient information of the power. In some of these places the upturned pines have lain so long that they are rotten throughout. There are the stones, too, large and small, which formed part of the destroying storm. Returning to George, I found him hard at work in the course of an avalanche, and I took a seat behind him for a while, using for that purpose both a pine and the stone which had over-turned it. Heard a good deal of murmuring thunder-ing noise, but whether of thunder or avalanches in the distance, or waterfalls, could not tell.

The baths of Leuk are 4,500 feet above the level of the sea. We went to see our room in the inn and wash, and then before dinner went to the baths to see the poor creatures bathing, for the time was on, being from two o'clock until five. Our house was the Maison Blanche. Just before this is an abundant natural source of hot water, arranged in a little masonry, at which people drink, and from which the water flows into the baths. We were forewarned that a certain square wooden building before us was the baths, by a great noise and outcry of human voices, issuing from it as from a penny theatre; and we were also told we might enter by certain doors and see the bathers; we did so, and entering, found ourselves on a raised plat-form going across the building, and on each side of and beneath us were two tanks, making four altogether, filled with water and human creatures. The tanks were like each other, and each was about five feet deep, and had benches round the four sides on which the bathers sat, so as to be immersed up to the neck in the clear fluid. The people had gowns on, and their heads, being safe as regarded the water, were handsomely decorated ; for the males wore gay caps embroidered, &c., and the females also had their appropriate head-gear, decked out with ribbons and flowers. It is very odd, but I cannot comfortably call them ladies and gentlemen, for neither the appearance, the grouping, nor the sounds seemed to partake of such character, and yet this is all mere matter of custom and understood usage. The heads of the people were about level with our feet, and as we looked down upon them, some of the gents seemed quite willing to talk and joke with us. I could not help thinking that, feeling they must form a ridiculous object to a new comer, they had tacitly agreed to make common cause against such, and by loud remarks, laughter, and jokes, to put him down as soon as possible. In the middle of some of the tanks were handsome nosegays, raised on central tables. Little floating tables were moving about, sustaining in some cases tea and coffee cups, in others ladies' work, in others toys, and most of the people had a little basket, with keys and a handkerchief, floating on these tables. Some of the people looked single and rather alone in thought—certainly not in company ; others were in groups ; one handsome-looking well moustached and well capped man, with two young ladies equally well capped (all the rest of the bodies was merged in gowns), seemed to form in one corner a very happy sentimental group, taking refreshment and whispering together.

There were about sixty men and twenty women, principally French and Italian. Many of them were practising an odd, idle sort of amusement, for sitting thus up to the neck in the water, they had found out that by putting three fingers of one hand so that they should be compressed by the other hand, whilst both palms were together, they could, holding the hands just or partly under the water, and alternately compressing and relaxing them, make a little jet d'eau, and the rivalry was who could make the most abundant jet, throw it furthest, and direct it most cleverly either here or there. Ladies and gentlemen both seemed very happy in this sort of occupation, which they pursue for from *two* to *nine* hours per day for weeks together. What good it may do their bodies I do not know, but it certainly must relax their minds. I can scarcely imagine a vigorous or strong-minded person submitting to it on any account.

We returned to the inn, dined, and then rambled down the valley to find *the ladders*, and after some queer rambling in a wood, between the overhanging precipices, which form the wall of this singular valley, and the torrent flowing in the bottom, found them. In fact, we found two boys going to them, and they showed off the ladders famously. On the top of the running precipice is a village called ——, which has established a communication with the baths for pedestrians down the side of the precipice by these ladders. We went through the wood on a rough path, rising along the inclined bank of stones, earth, and trees, and getting more and more above the river, which continually descended, until at last the path could go no farther; and after clambering six or eight feet over rocky points, the next step was on to a rough upright ladder, pinned to the rock by wooden hooks driven into the crevices. At the top of the first ladder was a cavern, dry and comfortable, having a beautiful view of the valley beneath. From the floor at the entrance of this cavern rose another ladder, going you could not tell where until at the top of it, and when there more upright scrambling took us to a third ladder. We scrambled up there. Could see five, but did not go farther, for we had to return, and George had to sketch this strange road as well as he could. Both men and women of the village use this path, and there is no other except by going many miles round. It is a very remarkable thing, and quite as curious in its way as the Gemmi Pass itself.

Returning, we had a magnificent piece of rosy sunlight on the topmost snow Alp : everything else being grey or dark. It was very fine. In bed by half-past nine o'clock.

Friday, 23rd.—All night long were we kept awake by the noises in the inn. Our fears that we should disturb the waiter too soon, by being up and breakfasting by five o'clock, were very unnecessary, for long before that not merely the inn, but the town was alive.

The bathers, whatever good they may do to themselves, will be the death of others. From the first hour of bed there were noises in the house, and to sleep in a room in one of these wooden buildings, and to have such noises, was as if one had determined to go to bed in a drum, whilst half a dozen boys had determined to have a game on the outside. Two Frenchmen were in the next room to us. ' Ah ! que je suis malade ! ' says one, all night, coughing continually, so as to give us perfect assurance that he was so. Poor fellow ! Up he was twice in the night with light in his room, as I could see under the door. Then at three o'clock the church struck up a vigorous ringing. I

am sure I do not know what for, unless it was for
prayers for the invalids who go into the baths soon
after. Immediately after that our neighbours the
Frenchmen began to get up and talk. One, the malade,
was lively, the other wished to snooze a little longer,
but his companion bore all before him, and after much
talking I was glad to hear his door open and he go
out. What for I could not tell, but there were now
hopes of rest for an hour. Hopes, indeed! and nothing
else. For in less than ten minutes back he came, talking
in his loudest key, in the utmost trouble and distress.
The baths were not ready!   The fires not lighted!
The place not warm!   The water not yet in the tanks!
and it half-past three o'clock!!! He would complain to
the maire, or the parson, or to his friend, and he ill all
night. Such treatment never man met with! and so he
grumbled away, thinking much of himself, and very
little of anybody that might be his neighbours, until
about four o'clock, when he went out and did not
return. I trust he was sopped and soaked to his perfect
satisfaction. About half-past three o'clock there was a
particular snapping noise which excited my attention :
away I went to the window and found it was the
crackling of three great wooden fires, made in different
stoves of the baths, opening to the air and intended to
warm the buildings. When at the window, the interest of the scene kept me there. As I said before, the
principal source of water was just before the house,
between it and the baths, and now, in the early morning
light, issued out from the various surrounding houses
and hotels the poor creatures cloaked up and coiffed,
or dressed otherwise very queerly, each bringing a
little basket containing their keys, their handkerchief,

and a tumbler, with other things. To the spring they ran and began to drink the hot water : some were going, others coming, whilst waiters were washing various articles there quite unknown to me, but I suppose of bathing importance, and such a mixture of men, women, poor, rich (I suppose), at such diversity of occupation I never saw in one spot before. As they drank, when they had taken enough, off they ran to the baths to take their place in the wonderful fluid. And so their days ran on and ran away.

By ten minutes to six o'clock we had breakfasted, packed up, and come outside the house, ready to start and part, for here George and I parted ; not a pleasant thing to either, though we hoped to meet again in a few days ; but he had to go further and sketch, and I had to return home, i.e. to my dear wife and Emma, by Sabbath day. My purpose was to make two days from here to Thun, sleeping one night at Fruchtigen or Kandersteg, as it might be. The morning was fine, and I took my solitary way up the Gemmi pass, arriving at the top by half-past seven o'clock, which was very well. Fine clouds at the top with sunshine and snow falling. Soon arrived at the lake or Daubensee, meeting myriads of sheep bleating and ringing their bells very pleasantly to my ears. The glaciers on the left between the summit and the Daubensee, showed the moraines on it very well. It then became cold and dull, and feeling rather solitary and a little melancholy, and thinking of poor George left to himself, I had half a mind to turn back and join him. What nonsense ! Passed the place where yesterday we placed some flat black stones on the snow. Though there has been a good deal of sun since, they had not sunk in, but had slid on the inclined surface of the snow, some about six inches, others more, others less. At thirty-five minutes past eight, reached the inn at Schwarenbach, there rested awhile and had some tea and a mutton chop. I left this place at fifty minutes past eight ; it was cloudy, and rain immediately began and continued to Kandersteg. It was no use stopping to wait for it, for who could tell when it would cease, and who could help me if I met them? I had an umbrella, so on I went. For two hours on the mountain-top it was strong rain, with the wind against me ; and as the path was soon a stream, my great fear was that my shoes would not hold out. On the road I passed many poor pedlars and peasants carrying goods, for such is the principal mode of conveying things to the baths of Leuk. At eleven o'clock I passed the inn at Kandersteg, where we had slept two nights before, but as I was thoroughly wet, and otherwise quite well, on I went, hoping that the weather would clear and become dry before I reached Fruchtigen. The weather now broke a little in these lower parts, and I saw many beautiful specimens of cloud-making and rain-making ; the falls, too, were beautiful. I arrived at Fruchtigen at one o'clock, having made twenty-eight miles since six o'clock this morning. But I still felt my clothes damp, and knowing the difficulty of getting quick accommodation, I resolved to dine, and then start or stay as I might feel inclined ; so with a little *eau de vie de Cognac*, and a very good dinner, and also the advantage of a rest of an hour and a half, I started again. It was then twenty minutes to three, and I had fine weather through the rest of the beautiful valley of the Kander. I reached home at Thun by twenty minutes after six, in far better condition

than I expected, and very glad to be there. After tea
I felt a little stiff, and only then felt conscious of one
small blister.

Now from Leuk baths to Schwarenbach is $9\frac{1}{2}$ miles.
From Schwarenbach to Kandersteg is $10\frac{1}{2}$ miles.
And from Kandersteg to Fruchtigen is 8 miles.
And from Fruchtigen to Thun is $16\frac{1}{2}$ miles.

Making altogether about forty-five miles. This I did
in $10\frac{1}{2}$ hours, excluding the two hours of rest; so that I
think my strength cannot be bad or my reasoning (?)
very insufficient. I would gladly give half this strength
for as much memory, but—what have I to do with
that? Be thankful.

(Here flowers from the top of the Gemmi Pass were
fastened into the journal with great skill and taste.)

Tuesday, 27th.—More pleasant rambles — fine —
George came back about two o'clock, quite hearty and
well. Now we shall think of a move, and really the
changing character of the *table d'hôte* and other things
make me in love with the thoughts of home. Dear
England, dear home! dear friends! I long to be in
and amongst them all; and where can I expect to be
more happy, or better off in anything? Dear home,
dear friends, what is all this moving, and bustle, and
whirl, and change worth compared to you?

Wednesday, 28th.—It is beautiful to see mingled together the elements, materials, and results of the forest:
thousands—nay, myriads—of seedlings of all kinds,
with grown trees, and then the monsters of the forest,
and at last their fallen decayed trunks, which though
dead are still the centre of life to many other vege-
tables, and to countless numbers of animals. The ants
appear to make great use of dead trees.

To Mrs. Faraday's sister, Miss Barnard, he wrote :—

'Interlaken: August 1, 1841.

'Dear Jenny,—A whole month has passed away since we left you and all our friends ; and though we have for so long been absent, and without those cheering words and pleasant little chats, which are and ought to be a comfort to one another, still we have had your letters, and you can hardly think how great their value is to us. I speak of you as all together, for I cannot tell how to separate you one from another ; bound up together, as I trust, in the one hope, and in faith and love which is in Jesus Christ, you seem to me to become more and more as one to us ; for, as we increase our distance from you, there appears to be a separation of the few, or I ought rather to say of the many, that are with you from the rest of the world, and a distinction which I feel to be the greater now that I am away. And you too, dear Jenny, are away in a manner ; for though it is not distance which separates you, yet it has pleased God in his dealings to bring weakness of body over you, and so to lessen your power to enjoy those privileges which are granted to us to keep alive in our hearts the knowledge and love of the Lord Jesus Christ. But we know that these his dealings with his people who are found waiting upon Him are all mercy, and are needful to rule their rebellious hearts to the obedience of Him, and to see in Him everything which is necessary for our rest. How anxious, to be sure, we are to do something ! Often it takes the form of going to his house, or of joining in his worship, or of working in the labour of love in the profession—things that his people, constrained by his love, will be always found observing : but when it pleases God to take from us these privileges, leaving us his Word, which is all-sufficient, how often shall we find, by the workings and thoughts of our hearts, that in these things we were trying to do something on our own account! And so we may be encouraged to hope that He is thus showing us what is within ourselves, that He may turn us again unto Christ and to Him alone. Now, my dear Jenny, I hope I have not written anything to make you dull. I have no such thought; I am rather hoping to cheer both you and myself by the thought of Divine goodness and mercy, which make salvation not of any worth or work of ours, or any goodness or strength or fitness that we have, but alone of his sovereign grace and mercy.

'Now I have told you no news—my companion and dear wife and partner in all things will tell you enough of that. We often talk of you and speak of our joint hope; and the separation that now is makes us think of another that will follow, and that cannot be much longer delayed, and of the joyful hope of meeting never to part in that heaven where there are many mansions, and where the Saviour is gone to prepare a place for his people.

. . . . . .

'Ever, dear Jane, your affectionate brother,
'M. FARADAY.'

The Swiss Journal continues :—

Monday, Aug. 2nd.—Interlaken. A fine pleasant-feeling day, and some very pleasant walks to the Pavilion and hills, from which the views were beautiful. The Jungfrau has been occasionally remarkably fine : in the morning particularly, covered with tiers of clouds, whilst the snow between them was beautifully distinct ; and in the evening showing a beautiful series of tints from the base to the summit, according to the proportion of light on the different parts. At one time the summit was beautifully bathed in golden light, whilst the middle part was quite blue, and the snow of its peculiar blue-green colour in the refts. Some of the glaciers are very distinct to us, and with the telescope I can see the refts and corrugations of the different parts, and the edges from which avalanches have fallen ; the Neisen is also very often most beautiful in our views over the lake of Thun.

.   .   .   .   .   .

They dry fruits here in the sun, as cherries, apples, pears, &c. : for this purpose they spread them out on boards surrounded by little raised ledges. These boards are blackened, that they may absorb the rays of the sun and become hot.

Clout-nail making goes on here rather considerably, and is a very neat and pretty operation to observe. I love a smith's shop and anything relating to smithery. My father was a smith.

To-morrow we are to arrange things for a real Alpine excursion.

.   .   .   .   .

Tuesday, 3rd.—We passed many waterfalls, and arriving at Lauterbrunnen, saw at once the Staubach, or principal waterfall of the neighbourhood. Our wives soon joined us, and then we went to the fall and looked about us. The water descends from an immense height, and is very beautiful ; but there is not much of it. If it falls clear of the rock, which depends on the wind, then it becomes rain long before it reaches the bottom, and varies very much in its appearance according to the light and position. In some positions the middle part disappears altogether; in others it looks well and full, resembling a slowly descending gauze veil, narrow, and long and perpendicular, or else waving with the wind, but always moving downwards, issuing slowly from above, and entering the ground on the rock on which it may fall. By watching a fold of the aqueous drapery one might judge of the time of descent, and it took forty-three beats of my watch, or nearly twenty seconds, to descend from the top to the bottom. We had three ordinary dirty-looking girls, who sang some Swiss music in parts very prettily and with good effect. We also had a cannon fired off (the usual toy) that we might hear the echoes.

.   .   .   .

Thursday, 5th.—Wengern Alp. A little further on was a boy with a mountain horn, by means of which he woke up the mountain echoes in an extraordinary manner. It was about six or seven feet long and was made of wood in two pieces, which had been hollowed out separately and then bound together by strips of the willow into one firm instrument. By dexterous blowing, the lad could bring out the harmonic notes of two or even three octaves, and so made his rough instrument discourse excellent music. It was rich, full, and very pleasant, filling these immense spaces with sound. A wall of rock a good way off returned a fine echo, the time being such that five or six notes were given back to us after the horn was silent; and as different parts of the precipice returned the sound at different times, very beautiful combinations of the notes took place—the distant faint echo of the echo lingering beautifully on the ear at last. He then gave us a bang with an iron cannon, but that was not so good. It should not be heard after the horn.

．　　．　　．　　．　　．　　．

We now heard an avalanche, and hastened our steps. We came into a wood—most picturesque. Pines were blown down and crossed our path, and we wondered how mules could ever pass along it. We came on to a rising ground on the top of a deep precipitous glen or chasm, and saw opposite to us the sources of the Silberhorn, the Jungfrau, and other wonderful summits which here rose before us, and sent down continually great streams of water rushing down in every form of fall, and every now and then thundering avalanches. The sound of these avalanches is exceedingly fine and solemn. It is the sound of thunder known to be caused by a fall of terrestrial matter, and conveys the idea of irresistible force. To the sight the avalanche is at this distance not terrible but beautiful. Rarely is it seen at the commencement, but the ear tells first of something strange happening, and then looking, the eye sees a falling cloud of snow, or else what was a moment before a cataract of water changed into a tumultuous and heavily waving rush of snow, ice, and fluid, which, as it descends through the air, looks like water thickened, but as it runs over the inclined surfaces of the heaps below, moves heavily like paste, stopping and going as the mass behind accumulates or is dispersed.

．　　．　　．　　．　　．　　．

Friday, 6th.—As we proceeded into the higher grounds at a height of 2,000 feet above Lauterbrunnen, the sun's rays proceeded in such a direction as to give us a fine rainbow, or at least part of one, in the waters of the Staubach and other falls. So that when we looked down from this great height upon the distant gauze-like films, they took all sorts of beautiful colours, varying, as we stepped backwards and forwards, over a space of fifteen or twenty yards.

We soon came to the Hotel de Jungfrau, and secured rooms for the night.

The avalanches were in plenty, and some very fine. The clouds gathered, and at last a distant storm with thunder and lightning came on, and a flickering gust of wind. Then a succession of exceedingly fine cloud effects came on, the blue sky appearing in places most strangely mixed with snow-peaks and the clouds. To my mind no scenery equals in grandeur the fine sky-effects of such an evening as this. We even had the rose-tint on the snow tops in the highest perfection for a short time.

Saturday, 7*th*.—We started for Grindelwald at ten o'clock. The snow-peaks were smoking abundantly —that is, the wind which passed over them formed with the air which crept up on the leeward side a cloud which continuously passed off from the peak some distance into the clear air. We soon passed the summit of our road at the Scheideck, and descending, came first into a forest of rugged blanched pines, dead from a very hard winter, but white as snow and shining beautifully in silvery desolation.

The world looks very bare here even at this season. Slate and granite occur. The streams are clear as ice of the glaciers, clear as crystal. Lower down we came to a couple of women sitting on the roadside and

making lace. They sang some Swiss songs very prettily;
but this has now happened so often and is so palpably
for money, and money only, that the sounds are losing
all their charms. The motive is too evidently self-
interest, and the self-interest is too evidently a great
degradation of the character of the people. One
regrets to acknowledge it, but begging is pursued in a
shameless manner. The mother working at a comfortable cottage door will send her child to beg of the
passers-by; the infant at the roadside that cannot yet
speak even its own language holds out its hand for
money. Give, give, is ever on the countenance.

Grindelwald upper glacier. The colour of this ice is
most beautiful, giving in the different fissures every tint
of blue, from the palest through Prussian blue to black.
The man took me into a low flat cavern. Its floor was
clear ice, beneath which was another similar cavern. Its
roof also was clear blue ice—its extent was thirty or forty
yards, but its height not more than five feet in the
highest place; and whilst standing on the floor we
could see through it, the waters running in the cavern
below. In melting from the contact of air, the under
surface generally takes a groined and concave form.
Thus many parts of this floor formed, as it were, a combination of rude plano-concave lenses, through which
the rivers of water below presented every shape and
size of cascade, rapid, &c. It is this kind of cavern that
gives origin to the glacier thunder, for as the thawing
continually proceeds—in summer, at least—the ice at
last becomes too weak to support such flat roofs, and
then they fall in tons and hundreds of tons at once.

I rambled about this glacier a long time, going up
the side to see the scratched rocks and lateral moraine.

The flowers here were beautiful and rich, glowing in
the sunshine of the fine day.

.   .   .   .   .   .

On returning we had tea, and after tea the weather
broke up; the evening sun shone out on the glorious
snow tops and fantastic clouds beneath them in the
most admirable manner. Nothing can surpass these
sunsets.

.   .   .   .   .   .

Tuesday, 12*th*.—Brienz Lake. George and I crossed
the lake in a boat to the Giessbach—he to draw and I to
saunter. The day was fine, but the wind against the
boat ; and these boats are so cumbrous, and at the same
time expose so much surface to the air, that we were
above two hours doing the two miles, with two men
and occasionally our own assistance at the oar. We
broke the oar-band ; we were blown back and sideways. We were drawn against the vertical rock in a
place where the lake is nearly 1,000 feet deep ; and I
might tell a true tale which would sound very serious,
yet after all there was nothing of any consequence but
delay : but such is the fallacy of description. We
reached the fall and found it in its grandeur ; for, as
much rain fell last night, there was perhaps half as
much more water than yesterday. This most beautiful
fall consists of a fine river which passes by successive
steps down a very deep precipice into the lake. In
some of these steps there is a clear leap of water of
100 feet or more, in others most beautiful combinations
of leap cataract and rapid, the finest rocks occurring
at the sides and bed of the torrent. In one part a
bridge passes over it. In another a cavern and a path
occur under it. To-day every fall was foaming from the

abundance of water, and the current of wind brought down by it was in some parts almost too strong to stand against. The sun shone brightly, and the rainbows seen from various points were very beautiful. One at the bottom of a fine but furious fall was very pleasant : there it remained motionless whilst the gusts and clouds of spray swept furiously across its place and were dashed against the rock. It looked like a spirit strong in faith and stedfast in the midst of the storm of passions sweeping across it, and though it might fade and revive, still it held on to the rock as in hope and giving hope, and the very drops which in the whirlwind of their fury seemed as if they would carry all away were made to revive it and give it greater beauty.

How often are the things we fear and esteem as troubles made to become blessings to those who are led to receive them with humility and patience.

In one part of the fall the effect of the current of air was very curious. The great mass of water fell into a foaming basin ; but some diverted portions struck the rock opposite the observer, and collecting left it at the various projecting parts. But instead of descending, these hundred little streams rushed upwards into the air, as if urged by a force the reverse of gravity ; and as there was little other spray in this part it did not at first occur to the mind that this must be the effect of a powerful current of air, which having been brought down by the water was returning up that face of the rock. To my mind this fall very much surpasses the Staubach in beauty, but it does not make a good subject for the artist.

Saturday, 14*th*.—The walking of this day was about

thirty-two miles, including the pass of the Brunig twice.
It was quite enough, and now and then we were both
rather warm.   Still health is excellent.

.   .   .   .   .   .

From the Lake of Brienz Mrs. Faraday wrote to
Mr. Magrath :—

MRS. FARADAY TO MR. MAGRATH.

'Dear Mr. Magrath,—Mr. Faraday seems very
unwilling to write letters ; he says it is quite a labour
to him, and that everyone advises that he should take
thorough rest ; and that he is quite inclined to do so.
I can certainly say nothing against all this, but I am
anxious that such an old friend as you are should not
be neglected altogether.   I will therefore take the
opportunity of his absence (he is exploring the pass
of the Brunig) to begin a letter for him, and to tell
you how we are going on. . . . We have been
absent from home six weeks now, which we consider
about half our time, and we have had upon the whole
favourable weather and seen a great deal of beautiful
scenery.   We expect to reach Lucerne in about a
week.   Any letter sent from England till the 25th of
this month may be directed there.   I think Mr. Young
will be quite satisfied with the way my husband em-
ploys his time.   He certainly enjoys the country exceedingly, and though at first he lamented our absence
from home and friends very much, he seems now to
be reconciled to it as a means of improving his general
health.   His strength is, however, very good ; he
thinks nothing of walking thirty miles in a day (and
very rough walking it is, you know), and one day he

walked forty-five, which I protested against his doing 1841.
again, though he was very little the worse for it. I Æt. 49.
think that is too much. What would Mr. Young say
to that? But the grand thing is rest and relaxation
of mind, which he is really taking. There are not so
many calls upon his memory here even to remember
people's names.

. . . . . .

Faraday finished the letter.

'Brientz: August 15, 1841.

'My dear Magrath,—Though my wife's letter will tell
you pretty well all about us, yet a few lines from an
old friend (though somewhat worn out) will not be
unpleasant to one who, like that friend, is a little the
worse for time and hard wear. However, if you jog
on as well as we do you will have no cause for grumb-
ling, by which I mean to say that I certainly have
not ; for the comforts that are given me, and above all
the continued kindness, affection, and forbearance of
friends towards me, are I think such as few experience.
And how are things with you? I must ask the question, whether I can hear the answer or not. Remember us most kindly to Mr. Young. We often have
to think of him for many reasons. I will give no
opinion at present as to the effect of his advice on
my health and memory, but I can have only one
feeling as to his kindness, and whatever I may forget, I think I shall not forget that. Amongst other
things, say that the net for the cloaks and coats is most
excellent, and has been several times admired for its
utility. It is droll to think what odd gatherings go
into it sometimes in a hurry. If you happen to see

Mr. Brande or Sir James South, remember me very kindly to them. Now as to the main point of this trip, i.e. the mental idleness, you can scarcely imagine how well I take to it, and what a luxury it is. The only fear I have is that when I return, friends will begin to think that I shall overshoot the mark ; for feeling that any such exertion is a strain upon that faculty which I cannot hide from myself is getting weaker, namely, memory, and feeling that the less exertion I make to use that, the better I am in health and head, so my desire is to remain indolent, mentally speaking, and to retreat from a position which should only be held by one who has the *power* as well as the will to be active. All this, however, may be left to clear itself up as the time proceeds ; and now farewell, dear Magrath, for the present, from your affectionate friend,

' M. FARADAY.'

Wednesday, 18*th.*—Fine day all day. We started at 7 o'clock, our wives on horseback, we on foot, and began to mount into the upper Haslithal over the extraordinary stone rock which as a dyke divides it from the lower. On we went, very wet in the feet from the extraordinary quantity of dew which filled every herb with gems, and as the sun shone surrounded us with continual rainbows on the ground.

Grimsel Hospice : Here I found Forbes, and also found that Agassiz had been there and was gone up to his hut on the Grimsel glacier, intending to sleep there tonight. Seven or eight have been passing day and night in this hut, consisting of slabs of ice and stones, on the moraines of the glacier, for a fortnight past. I arranged for George and I to go up with Forbes to the glacier

to-morrow. It is four hours there and four back. The
hut has descended with the glacier 1,000 feet since last
year at this time.

Tuesday, 19th.—Not well this morning, and cold increased rather. So as the day was fine we altered our
minds and set off on our journey for the Rhone glacier.

.   .   .   .   .   .

We set off again for the pass of the Furca and for
Hospenthal. Our ascent to the Furca led up by the side
of the fine glacier from which the river Rhone originates,
and amongst a full spring development of flowers and
buds. Many plants were here just springing out of the
ground, and at the top we had to descend immediately ;
and as the snow had not yet left the paths open, we
had to slide down long banks of snow, filling up the
bottom of the Alpine crevice hereafter to form a valley.
Under these banks many a young river ran. Our
wives dismounted and travelled well. Rapidly it could
not but be, but it was well also. When we gained the
uncovered path it was very bad and sterile, but we
were cheered at last by a distant sight of Hospenthal,
the place on the great St. Bernard pass which we were
going to. Evening now crept on ; we came to a village ;
every child in it begged, and then hooted ; and there can
be no better illustration of the harm that indiscriminate
giving occasions than in this country. The character
of the whole population has been seriously lowered and
injured by it. Darkness came on : our path was a
mere rough footway, and the horses being fatigued
began to stumble more than before over the stones.
The valley now spread out, but as a consequence many
bogs were formed through which we had to feel our
way. These rivers from the mountain sides had strewn

wide areas with rocks and stones, amongst which the waters ran ; and when, crossing these, we came to the present narrow course of the stream, George and I had to cross by a plank, whilst our wives had to trust their horses in fords of most awkward appearance, and sometimes on the very edge of falls several feet in height. Even the horses at times hesitated to enter among the large unseen stones and the roaring waters. Four different torrents of this kind had we to pass. There was no moon and but little star-light, but fortunately the granite here is of a light colour, and so the difference between the bare road and the green sides was our guide. At one place the path was covered with water and sand for some distance, and amongst these streams and rivers we had to find our way. Our guide and the horseman here certainly behaved well. They picked their way well when I could scarcely see the tail of the horse behind which I walked, and when things looked very dull cheered themselves, us, and even the horses, by singing some Alpine songs. At last we arrived at the village of Hospenthal, greatly to the surprise of the landlord of the inn, for it was half-past nine. We were well received ; supped, went to bed, and dreamed.

Monday, 23rd.—Arrived at Lucerne.

Wednesday, September 1st.—Lucerne. At last the month is come in which I hope to see home again. My cold is heavy.

Friday, 3rd.—At last had the doctor, and he directed a warm bath and perspiration, but no medicine. The walk out to the bath was quite delightful. I felt as if I had liberty again, and could enjoy the beauties of nature. It was like an escape from prison.

Friday, 10*th*.—Zurich. The day was beautiful, and
we visited first the school for the blind, and also
the deaf and dumb. The deaf and dumb were truly
astonishing. To hear them speak, and yet to know
that they cannot hear themselves. To perceive that
when they speak to us jointly with each other, we all
know the words but by different senses, since they
perceive by seeing what we know only by hearing, is
very wonderful. I spoke a word or two to one of
them, but spoke it silently, making the motions of the
mouth, but using no voice. My companions did not
know what I said, but the deaf boy knew; and he who
had been born dumb repeated the words which those
possessing all their senses had no knowledge of. This
teaching of the deaf and dumb is a thing kindly, wonderfully, and well done. In the course of the day we
walked to the Katz bastion, a part of the fortifications,
now converted into a beautiful garden, affording a fine
view of the town and neighbourhood. It is a curious
and an amusing consideration which arises on comparing
these people of Zurich with the Parisians. These are
busy in razing all their fortifications; those in erecting
such works round their city of Paris. A few years will
suffice to show which has the most wisdom. Probably
the two steps are equally wise for the respective people
by whom they are taken.

Saturday, 11*th*.—Another fine day; I delight to
notice these fine days as a set-off to the many dull
reports we have had. Went in an omnibus to Schaffhausen: about thirty miles off. We stopped at the
Rhine falls whilst George went on to the inn and
returned to us, and we then enjoyed these fine falls,
passing to both sides of the river for the purpose. A

1841.

Æt. 50.

new gallery has been built close to, and over, the fall. We were there in the finest lights of evening, whilst the sun still formed rainbows in all parts of the mist, and the union of sublimity and beauty in the desperately bounding waters is extraordinary. I think no fall could surpass it in this effect from the spot where we stood ; for if the river were as large as that of the Niagara Falls, still one is here so thoroughly in the midst of the rush, that the eye can see nothing much beyond twenty feet of distance ; but within that space all that water can present of power and beauty is shown. When the shades of evening came on we went to Schaffhausen.

Monday, 13*th*.—Went again with George to the falls of the Rhine, and picked up a flower in remembrance. They are very beautiful, but there is nothing like the beauty on them under the midday sun there was in the evening lights of Saturday.

Friday, Sept. 17*th*.—We reached Heilbron about 9 o'clock P.M., and slept there. Entering it by dark night, we could see nothing scarcely ; but when in our room we saw a light, rather steady in degree, moving about up in the sky over a small space. Upon inquiry we found it was a watchman, placed in the highest gallery of the cathedral tower. There he walked all night, watching over the city, and showing his lanthorn to give proof of his watchfulness. He had a trumpet on which at the half-hour he blew a single note, or sometimes repeated it. But if he saw signs of fire or other evil to the city, then he blew an alarm. It reminded one strongly of the prophetic figure of the watchman, and of his certain or uncertain note.

Monday, 20*th*.—Manheim. Up at 5 o'clock, and off by

6 o'clock, in the steamboat, intending to proceed down the Rhine to Coblentz, which place we reached by 4 o'clock. The day was fine, but one cannot think much of the Rhine after leaving Switzerland. Besides, the time of day did not give good shadows, and the vineyards were particularly ugly. Even their greenness, which is the only good character about them, is passing into dull brown. The ruined castles were as beautiful as ever; but these beauties will not compete with those of mountains, valleys, rocks, and varied wood; and the Rhine hills have no grandeur.

Thursday, 23rd.—The Moselle. No steamboat, no carriage, no practicable road except for horses; and surpassing all in retarding influence, such torpor in the people as I never saw before in the most out of the way place. At last we procured an open paddle boat, and getting in with our luggage, we pushed out into the river, and floated down with the stream, a man and a boy helping by paddling. At first it was fine, then a storm came on. Then the man put the boy on shore, and sticking up an oar or paddle for a sail, went himself to the stern to steer and paddle. Then I took the paddle and dabbled awhile. Next we picked up a man, who took my place, and at last we arrived at Coblentz again, with a fine evening, and gave up the Moselle.

Wednesday, 29th.—About 1 o'clock this morning we were woke up by the alarm of fire—always an awful alarm at sea. Most of us laid quiet, hoping it was nothing, and anxious, I suppose, not to increase the fear and tumult consequent on the thought. Gradually all was set right, but afterwards I was told there was some ground for the fear; perhaps it was small. Then

1841.

Æt. 50.

1842-43. the wind rose in force, and, being contrary, kept us
Ær.50–52. long on the sea. At last it was a heavy gale that blew,
and I believe nearly all were ill on board. I did not
leave my berth till about 7 o'clock, and then found the
deck wet, the sea working, but the sky bright. The
wind was very powerful, still steam sped our way.
The journal ends thus :

Crossing the new London Bridge street we saw M.'s
pleasant face, and shook hands, and though we separated in a moment or two, still we feel and know we
are where we ought to be—at home.

In 1842 and 1843 but little original research was
done. Many lectures, however, were given for the
good of the Royal Institution, and there was some
work for the Trinity House. His reputation showed
itself by the Prussian order of merit, and by other
titles which he received, and by a very remarkable
letter from Prince Louis Napoleon, at that time a
prisoner at Ham. His letters are not of much interest.

I.

In 1842, on June 1, tempted probably by the
remarkable discovery which Sir W. Armstrong had
published two years previously, he made the first
note of fresh experimental researches in electricity,
after twenty months of rest. It was on the electricity
of steam, 'to see whether it might not be from friction
against metal ; as the metal cock or pipe.' He continued these experiments throughout June, November,
December ; and January 26, he sent the eighteenth

series of 'Researches on the Electricity evolved by the 1842–43.
Friction of Water and Steam against other Bodies' to Æt.50–52.
the Royal Society. Faraday showed that 'the cause
of the evolution of electricity by the liberation of
confined steam is not evaporation ; and further being,
I believe, friction, it has no effect in producing, and is
not connected with, the general electricity of the atmosphere.' The origin of the electricity was uncertain
until these researches were made.

In September he made some experiments on the
change of water and ice into each other.

To this subject he returned in 1850, when he gave
a lecture on it at the Institution.

In 1843, when the paper on the electricity evolved
by steam was ended, there was no more work in the
laboratory. Indeed, scarcely any note was made
until the end of May 1844, that is, he took nearly
eighteen months more of rest from experimental research. During this time he lectured for the Institution, and he did his work for the Trinity House. But
excepting his paper on the electricity evolved by
steam, no laboratory work was done for upwards of
three years.

For the Institution, in 1842, he gave two Friday
evening lectures, one on the lateral discharge in lightning rods, and another on the principles and practice
of Hullmandel's lithographic printing. In 1843, after
Easter, he gave eight lectures on electricity, from the
same notes as he used in 1838, and with the same
experiments. He gave three Friday discourses on
some phenomena of static electric induction ; on
lighthouse ventilation ; at this lecture he read a letter

from the keeper of St. Catherine's light, dated
February 19, 1843 :—

'The weather to-day forms a comparison of February 1841 ; but, pleasing to say, no damp. Sir, your plan has driven the enemy out. I entertain now not the slightest fears of him ever coming again to cause such labour as you witnessed on February 4, 1841.'

The third Friday discourse was on the electricity of a steam current.

In 1842 he gave the Christmas Lectures on electricity.

In March, May, August and September, he made different reports to the Trinity House.

In August he was at Newcastle, examining Cookson's operations for grinding the glass for lenses.

In 1843, for the Trinity House, he went to the South Foreland lighthouses regarding their ventilation. He inspected the dioptric light of the first order, which had just been constructed in France and put up there by French workmen, and compared its consumption of oil with the fifteen Argand burners which were previously in use.

He sent to the Philosophical Magazine ' a paper on static electrical inductive action. Among his notes the following occurs :—' Propose to send to the " Phil. Mag." for consideration the subject of a bar, or circular, or spherical magnet—first, in the strong magnetic field ; then charged by it ; and, finally, taken away and placed in space. Inquire the disposition of the dual force, the open or the related powers of the poles exernally, and if they can exist unrelated.   The difference between the state of the power, when related and when not, consistent with the conservation of force. Avoid any particular language. Should not

pledge myself to answer any particular observations, 1842–43.
or to any one, against open consideration of the sub-Æt.50–52.
ject. Want to direct the thoughts of all upon the
subject, and to tie them there; and especially to
gather for myself thought on the point of relation or
non-relation of the antithetical force or polarities.'

He also sent a paper to the ' Philosophical Magazine '
on static electrical inductive action, and on the
chemical and contact theories of the voltaic battery.

The ventilation of lighthouses led him to apply
the same principle to the ventilation of oil and gas
lamps. He gave the invention to his brother. Thus,
after describing the invention, he says :

' December 10, 1842.

' And now, dear brother, believing this particular
arrangement of the ventilating flue to be my own
invention, and having no intention of turning it to any
pecuniary use for myself, I am most happy to give
freely all my rights in it over to you, or any body
you may name for your good ; and as Mr. Carpmael
says we may legally and equitably make this transfer
of rights in this way, I write to you this letter describing the principle and arrangement of the invention, as far as I have carried it. Hoping it may be
productive of some good to you, and of no harm or
trouble,

    ' I am, my dear Robert, your affectionate brother,
' M. FARADAY.'

At a meeting of the Civil Engineers, June 13, 1843,
he had a paper on the ventilation of lamp burners.

M 2

II.

In 1842 he was made Chevalier of the Prussian Order of Merit (one of thirty), and Foreign Associate of the Royal Academy of Sciences, Berlin.

In 1843 he was made Honorary Member of the Literary and Philosophical Society of Manchester, and Useful Knowledge Society, Aix-la-Chapelle.

He received a medal from the Royal Society of Copenhagen, for which he thanks Professor Oersted in the following letter :—

PROF. OERSTED TO FARADAY

'My dear Sir,—It is with very great pleasure that I avail myself of an opportunity of expressing my respect for and strong remembrance of you, both for your work's sake, and for the personal experience I have had of your kindness.

'May you long live to advance as you have done, and to make your friends happy. I send you papers now and then as slight tokens of my respect, and hope you will accept them in good part.

'I have lately received from the Royal Society of Copenhagen the medal struck in 1842, and esteem it as a great favour and honour. Will you do me the kindness to return my very grateful thanks for such remembrance of me on the part of the society, to which I owe a deep debt of gratitude for its approbation of my exertions eleven years ago, and which was to me of great value and encouragement.

'I am, my dear friend, your obliged and grateful 1842–43.
servant,

ÆT.50–52.

'M. FARADAY.'

PRINCE LOUIS NAPOLEON (ÆT. 35) TO FARADAY.

'Fort of Ham : May 23, 1843.

'Dear Sir,—You are not aware, I am sure, that since I have been here no person has afforded me more consolation than yourself. It is indeed in studying the great discoveries which science is indebted to you for, that I render my captivity less sad, and make time flow with rapidity.

'I submit to your judgment and indulgence a theory of my own on voltaic electricity, which was the subject of a letter from me to M. Arago on April 23 last, and which I here subjoin. M. Arago was kind enough to read it to the Academy, but I do not yet know the general opinion on it. Will you have the kindness to tell me sincerely if my theory is good or not, as nobody is a better judge than yourself?

'Permit me also to ask you another question that interests me much, on account of a work I intend soon to publish : *What is the most simple* combination to give to a voltaic battery, in order to produce a spark capable of setting fire to powder under water or under ground? Up to the present I have only seen employed to that purpose piles of thirty to forty pairs constructed on Dr. Wollaston's principles. They are very large and inconvenient for field service. Could not the same effect be produced by two spiral pairs only, and if so what can be their smallest dimension?

'It is with infinite pleasure that I profit of this opportunity to recall myself to your remembrance, and to assure you that no one entertains a higher opinion of your scientific genius than, yours truly,

'NAPOLEON LOUIS BONAPARTE.

' I beg to be kindly remembered to Sir James South and to Mr. Babbage.'

The letter to M. Arago exists among the papers of Faraday. His answer to it has not been yet found.

PRINCE NAPOLEON TO M. ARAGO.

' Monsieur,—Toutes les fois qu'on trouve, ou qu'on croit avoir trouvé, quelque chose de nouveau qui intéresse la science, c'est à vous qu'on s'adresse ; car vous êtes l'axe autour duquel se meut le monde scientifique, et on est persuadé si l'on a raison de mériter votre approbation, comme si l'on s'est trompé de pouvoir compter sur votre indulgence. L'idée que je vous soumets aujourd'hui est relative à une nouvelle théorie de la pile voltaïque.

' La source de l'électricité a été attribuée par Volta au contact de deux métaux dissemblables. Davy a partagé cette opinion, mais depuis, de célèbres savants, et entre autres l'illustre Faraday, ont attribué à la décomposition chimique du métal la seule cause de l'électricité. Adoptant cette dernière hypothèse, je me suis dit : Comme dans la pile il n'y a jamais qu'un seul métal qui soit oxidé, si l'électricité n'est due qu'à l'action chimique, le second métal ne doit jouer dans cet accouplement qu'un rôle secondaire. Quel est ce rôle ? C'est, je crois, d'attirer ou de conduire l'électricité développée par l'action chimique d'une manière analogue à ce que se passe dans la machine électrique ordinaire. En effet, dans celle-ci l'électricité dégagée par le frottement traverse un milieu conducteur imparfait, qui est l'air, et est attirée et conduite par un parfait conducteur, qui est le métal. Dans la pile l'électricité produite par l'oxidation d'un métal quelconque traverse un milieu conducteur imparfait, qui est le liquide, et est recueillie et transmise par un parfait conducteur, qui est le métal adjacent.

' Cette idée m'ayant parue claire et simple, je cherchai le moyen d'en prouver l'exactitude par l'expérience, et je fis cet autre raisonnement : S'il est vrai que des deux métaux employés dans la pile l'un ne serve que de conducteur, on pourra le remplacer par le même métal que celui qui s'oxide, pourvu qu'il soit plongé dans un liquide qui permette à l'électricité de passer sans attaquer le métal.

'L'expérience vint confirmer mes prévisions. Je construisis deux couples suivant le principe de Daniell, mais avec un seul métal. Je plongeai un cylindre de cuivre dans un liquide composé d'eau et d'acide nitrique, et le tout contenu dans un tube de terre poreuse, et j'entourai ce tube d'un autre cylindre en cuivre, plongeant dans de l'eau acidulée avec de l'acide sulfurique, mélange qui n'attaque pas le cuivre. Ayant établi les communications comme on le pratique ordinairement, je décomposai facilement avec cette pile de deux couples de l'iodure de potassium dissout, et ayant adapté aux extrémités des deux pôles deux plaques de cuivre, plongeant dans une dissolution de sulfate du même métal, je recueillis au pôle qui était en rapport avec le cuivre attaqué un dépôt de cuivre.

' Je fis une seconde expérience avec du zinc seul.

Je mis dans l'auge poreux du zinc et de l'eau acidulée avec de l'acide sulfurique, et j'entourai ce tube d'un autre cylindre en zinc, plongeant dans de l'eau pure tiède. Avec deux couples semblables ainsi formés je décomposai également de l'iodide de potassium, et j'obtins, en prenant des précautions nécessaires, un dépôt de cuivre au pôle qui était en rapport avec le zinc attaqué.

'Enfin je renversai l'ordre habituel des métaux, et mis du cuivre dans le centre du tube, plongeant dans de l'eau et de l'acide nitrique, et j'entourai ce tube d'un cylindre en zinc, plongeant dans de l'eau pure, et j'obtins encore ainsi une pile assez forte.

'J'aurais voulu pouvoir mesurer avec soin la différente force des courants électriques, mais il m'a été impossible de le faire, faute d'un galvanomètre. Mes efforts pour en construire un n'ont pas réussi, parce que les aiguilles aimantées furent toujours déviées par l'attraction des barreaux de fer qui entourent mes fenêtres.

'Cependant, d'après les expériences que j'ai pu faire, il me semble démontré que dans la pile la cause de l'électricité est purement chimique, puisqu'il suffit d'un seul métal pour produire un courant, que le métal qui n'est pas oxidé ne fait que transmettre l'électricité comme dans l'électricité ordinaire. Enfin, que chaque métal est positif ou négatif, anodé ou cathodé à lui-même ou à d'autres, suivant le liquide dans lequel on le plonge.

'Je vous transmets ces détails avec une grande réserve, car je n'ai point fait de la chimie ni de la physique mon étude spéciale, et c'est seulement l'hiver dernier que, pour abréger les heures de ma captivité,

je me suis livré à quelques expériences en étudiant, 1843.
avec le plus vif intérêt, les travaux des hommes illustres Æt.51–52.
qui font faire tant de progrès à la science.'

Among the correspondence of Faraday, two other
notes from Prince Louis Napoleon exist; they are both
undated, and the answers of Faraday are not to be
found.

PRINCE L. NAPOLEON TO FARADAY.

' The Prince Napoleon presents his compliments to
Mr. Faraday, and begs him to have the kindness to
answer to a metallurgique question which is for the
Prince of rather great importance.

' The Prince should be very anxious to find one alloy
which would be *less fusible* than lead, and at the same
time *nearly as soft.*

'The Prince thinks that lead and zinc (mixed together) would perhaps answer the purpose, but having
no means to make experiments, the Prince would be
extremely obliged to Mr. Faraday if he could appoint
any person to make these trials. The Prince would with
pleasure pay all expenses.

' The Prince is very sorry to give so much trouble
to Mr. Faraday, but he relies upon his kindness.'

PRINCE L. NAPOLEON TO FARADAY.

' Le prince Napoléon fait ses compliments à M. Faraday, et le remercie des renseignements qu'il lui a
transmis il y a quelques semaines. Il espère que ce
n'est pas abuser de sa complaisance que de lui adresser
aujourd'hui la question suivante : Quelle est la plus

1842–43. grande dimension que l'on peut donner à la mousse de
Æᴛ.50–52. platine (cette préparation chimique du métal qui a la
propriété d'inflammer l'hydrogène)? Le prince sera
très-obligé à M. Faraday s'il veut bien lui répondre
un de ces jours quand il n'aura rien de mieux à faire,
car il serait désolé de le détourner, même pour un
moment, de ses occupations, qui sont si importantes
pour la science.

'1 Carlton Gardens, Friday.'

III.

During these two years a few letters show Faraday's
character; amongst them is one to Matteucci showing his great depression; and another to Dr. S. M.
Brown, who had asserted the isomerism of carbon and
silicon, and had had his experiments almost universally
rejected. He wished Faraday to witness an attempt to
transmute carbon into silicion, ' on the simple condition
of giving me a written testimonial to be used as
the Edinburgh Royal Society think fit.' Mr. Brown
also complained that he had called and had not been
admitted.

FARADAY TO DR. T. M. BROWN.

'Royal Institution: December 26, 1842.

'Dear Sir,—That which made me inaccessible to
you makes me so, in a very great degree, to all my
friends, *ill health connected with my head;* and I have
been obliged (and am still) to lay by nearly all my
own pursuits, and to deny myself the pleasure of
society either in seeing myself in my friends' houses, or
them here. This alone would prevent me from acced-

ing to your request.   I should, if I assented, do it 1842–43.
against the strict advice of my friends, medical and Æt.50–52.
social.

' The matter of your request makes me add a remark
or two which I hope you will excuse. Anyone who does
what you ask of me, i.e. certify if the experiment is
successful, is bound without escape to certify and
publish also *if it fail*, and I think you may consider
that very few persons would be willing to do this.   I
certainly would not put myself in such a most unpleasant condition.

' Again, why not test the experiment in Scotland, for
there you have published it ?   If Professor Christison
has given you letters, let him be your companion in an
experiment, and, if he likes, tell the world his judgment
on the matter.   His character is such that if you satisfy
him, and he conjoins his testimony with yours, I should
think you would not have much to fear as to the truth
of the discovery.

' I am, my dear Sir, very truly yours,

' M. FARADAY.'

To his friend Mr. Grove, who was at that time doing
his utmost to improve the Royal Society, Faraday thus
wrote, declining to take any active part in the work.

FARADAY TO W. R. GROVE, ESQ.

'Royal Institution : December 21, 1842.

' My dear Grove,—

.   .   .   .   .   .

As to the Royal Society, you know my feeling towards
it is for what it has been and I hope may be.   Its
present state is not wholesome.   You are aware that I

am not on the council, and have not been for years,
and have been to no meeting there for years; but
I do hope for better times. I do not wonder at your
feeling—all I meant to express was a wish that its circumstances and character should improve, and that it
should again become a desirable reunion of *all* really
scientific men. It has done much, is now doing much,
in some parts of science, as its magnetic observations
show, and I hope will some day become altogether
healthy.

> 'Ever, my dear Grove, yours sincerely,
> 'M. FARADAY.'

FARADAY TO C. MATTEUCCI.

'Royal Institution: February 18, 1843.

' My dear Matteucci,—I received your letter yesterday, and am much affected by your very kind inquiries
after one who feels as if his purpose of life in this
world were, as regards the world, passed, for every
letter of yours finds me withdrawn more and more
from its connections. My health and spirits are good
but my memory is gone, and it, like deafness, makes a
man retreat into himself.

.   .   .   .   .   .

' I think you are aware that I have not attended at
the Royal Society, either meetings or council, for some
years. Ill health is one reason, and another that I do
not like the present constitution of it, and want to
restrict it to scientific men. As these my opinions are
not acceptable, I have withdrawn from any management in it (still sending scientific communications if I

discover anything I think worthy). This of course
deprives me of power there.

. . . . . .

' With earnest congratulations to you on your last papers,

' I am, my dear Matteucci, your faithful friend,

' M. FARADAY.'

I.

The year 1844 is not remarkable for original research. It is noticeable chiefly for a speculation on the nature of matter, and it was followed by an opposite speculation on the physical lines of force. These two theories are of great interest, as they mark the imaginative part of the mind of Faraday, and show the way in which he let it act when he thought it was ' time to speculate.'

On May 23 he began to experiment upon one of his earliest subjects, the condensation of the gases. He added to pressure great cold, and he hoped to get fluid or solid hydrogen, nitrogen, and oxygen. He began with cyanogen: he worked in June, July, August, September, November, and December, and December 19 he sent a paper to the Royal Society. He failed in his object of solidifying oxygen or hydrogen, but he reduced six substances usually gaseous to the liquid state; and seven, including ammonia, nitrous oxide, and sulphuretted hydrogen, he made solid.

He sent to the 'Philosophical Magazine' a speculation touching electric conduction and the nature of matter. Elsewhere he calls this ' a speculation respecting that view of the nature of matter which considers its ultimate atoms as centres of force, and not as so many little bodies surrounded by forces, the bodies being considered in the abstract as independent of the forces, and capable of existing without them. In the latter view these little particles have a definite form and a certain limited size. In the former view such is not the case, for that which represents size may be considered as extending to any distance to which the lines of force of the particle extend. The particle, indeed, is supposed to exist only by these forces, and where they are it is.'

At the Institution he gave eight lectures after Easter on the phenomena and philosophy of heat. He ended this course thus : ' We know nothing about matter but its forces—nothing in the creation but the effect of these forces ; further our sensations and perceptions are not fitted to carry us ; all the rest, which we may conceive we know, is only imagination.' He gave two Friday discourses : the first on the nature of matter, the other on recent improvements in the silvering of mirrors.

His notes of the first lecture begin thus :—' Speculations, dangerous temptations ; generally avoid them ; but a time to speculate as well as to refrain, all depends upon the temper of the mind. I was led to consider the nature of space in relation to electric conduction, and so of matter, i.e. whether *continuous* or consisting of *particles with intervening space*, according to its supposed constitution. Consider this point, *remarking the assumptions everywhere.*

    .   .   .   .   .   .

' *Chemical considerations* abundant, but almost all *assumption*. Easy to speak of atomic proportions, multiple proportions, isomeric and isomorphic phenomena and compound bases ; and to account for effects we have only to hang on to assumed atoms the properties or arrangement of properties assumed to be sufficient for the purpose. But the fundamental and main facts are expressed by the term *definite proportion*,—the rest, including the atomic notion, is assumption.

' The view that physical chemistry necessarily takes of atoms is now very large and complicated ; first many elementary atoms—next compound and complicated atoms. System within system, like the starry heavens, *may be right*—but may *be all wrong*. Thus see how little of general theory of matter is known as fact, and how *much* is assumption.

. . . . . .

' Final brooding impression, that particles are only centres of force ; that the force or forces constitute the matter ; that therefore there is no space between the particles distinct from the particles of matter ; that they touch each other just as much in gases as in liquids or solids ; and that they are materially penetrable, probably even to their very centres. That, for instance, water is not two particles of oxygen side by side, but two spheres of power mutually penetrated, and the centres even coinciding.

' As I begin by a warning against *speculation*, so end by a warning against too much *assurance*. What is the experience to us of past ages—all *sure* in their days except the most wise—yet how little remains, and are we wiser in our generation ? Was earth, air, fire and water right ; then salt, sulphur, and mercury ; then phlogiston ; then oxyacids and oxygen ; now atoms ? We may be *sure* of facts, but our interpretation of facts we should doubt. He is the wisest philosopher who holds his theory with some doubt ; who is able to proportion his judgment and confidence to the value of the evidence set before him, taking a fact for a fact, and a supposition for a supposition; as much as possible keeping his mind free from all source of prejudice, or where he cannot do this (as in the case of a theory) remembering that such a source is there.'

Two remarkable letters were written to him regarding this lecture by two medical men, Dr. Mayo and Joseph Henry Green. They both looked at the lecture from a metaphysical point of view, and came to very opposite conclusions.

DR. MAYO TO FARADAY.

'56 Wimpole Street: March 6, 1844.

' My dear Sir, — You will, I trust, excuse my troubling you with some remarks on the admirable lecture, of which you have kindly favoured me with an abstract. Believing that no analytical inquiry has ever been set on foot, without some preconceived hypothesis, I imagine also that theory and hypothesis never need interfere with the *prosecution* of an inquiry. . . Your discoveries, indeed, sufficiently show the value of hypothesis. For no man uses its language more successfully than you do, as the associating agent in your analytical inquiries. In this respect your intellectual operations supply a striking proof of the value of a vivid imagination in a philosopher.

But I would suggest to you the following doubts as to the hypothetical expression which you are disposed to substitute for that, at present in use, of the atomic doctrine.

'Your atmosphere of force, grouped round a mathematical point, is not, as other hypothetical expressions have been in the course of your researches, an expression linking together admitted phenomena, but rather superseding the material phenomena which it pretends to explain. It resolves, in fact, as it would appear to me, all matter into a metaphysical abstraction. For it must all consist of the mathematical point, and the atmosphere of force grouped around it.

'You ought perhaps to carry your disposition to limit our real knowledge of things to effects and laws a little further, and apply it also to your own hypothesis. A mathematical point with an atmosphere of force around it, is in respect to the atmosphere of force an expression of certain effects. But what is the mathematical point?

'The question which the philosopher has to answer in deciding whether he should accept this or any other hypothesis on the subject, is whether it best interprets phenomena or is least at variance with them; the objection which you take to atoms on the ground of their uncertain magnitude is one which presumes that we pretend to more knowledge of them than those who entertain that theory *need* affect to possess. Indeed, your mathematical point is either a simple negation, as having neither magnitude nor parts; or is itself, after all, a material atom.

'The objection that *silver must vanish if its forces are abstracted* may prove the necessity of forces to our conception of silver, but does not disprove the necessity of silver to our conception of its forces; all that we can positively assert as known, are effects or forces; but we are organised and irresistibly impelled to assume substantiæ of which these are properties. Berkeley permitted himself to philosophise in regard to the external world, just as if he had not proved that our sensations are all that we can confidently assert as known to us.

.   .   .   .   .   .

'But I will not detain you longer. If my imperfect acquaintance with this class of subjects has occasioned me to write nonsense, pray tear my letter, but

'Believe me, your sincere friend and admirer,

'THOMAS MAYO.'

JOS. HENRY GREEN TO FARADAY.

'Hadley, near Barnet: July 3, 1845.

'Dear Sir,—I have read your lecture on the nature of matter with all the delight which any one must feel in finding the opinions which he has long held so ably vindicated and so clearly illustrated.

'There is, however, one difficulty which will be felt in adopting the theory of Boscovich, namely, that matter, or the physical agent, fills indeed space, that is by virtue of its forces, but does not *occupy* it. The ideal points which are the foci of forces attractive and repulsive, do not present any intelligible conditions for the origination and renewal of the forces.

'There is a want of the idea of *substance*. This, it is true, is unfortunately an equivocal word: but I flatter myself that in the appendix to the "Vital Dynamics," of which I ordered a copy to be sent to you,—namely, in the "evolution of the idea of power," I have given it a correct philosophical import in assigning to it an equivalent meaning with the "subject," id quod jacet *sub*,—that it is therefore essentially supersensuous, beyond the possible apprehension of the senses, but necessarily inferred as quomodo ejusdem generis, with that which constitutes our own subjectivity, and consciously known as will, spirit, power.

'This seems to me to be the true ground and key of all dynamic philosophy; but it has led me further, and I cannot but think that you have been also induced to extend your views in the same direction. Taught by your researches, that chemical combination depends upon the equilibrium and neutralisation of opposite forces, the liberation of which by decomposition resolves them into voltaic currents, I have been unavoidably forced back upon the question : If the electric forces are the true agents of chemical change, what share have the material substances or chemical stuffs in the phenomena ? And though my knowledge of the subject is too imperfect to permit me to come to any satisfactory conclusion, I must say that all the arguments I can muster bring me to the result, that these supposed stuffs are but the sensuous signs and symbols of the forces engaged in their production. Would that it were my good fortune to communicate with you more at large on this matter.

'Your obliged,
'Jos. Henry Green.'

This year for the Trinity House he only examined different kinds of cottons for the lamps of the lighthouses.

Some of his notes of the Haswell Colliery accident show how he worked at whatever he undertook.

The accident happened Saturday, September 28,
1844, about three o'clock P.M.

Tuesday, October 8th.—Went with Mr. Lyell from London to Durham.

Wednesday, 9th.—We went to Haswell and were at the inquest all day. This is the fourth day, and is an adjournment from last Wednesday. William Chiltern examined, pitman, Haswell Colliery; lampkeeper; had four Davy lamps brought to him; does not know the day; knows the numbers of the lamps; ninety-four belongs to Hans Ward, ninety-one to Thomas Turnbull, ninety-five to Mathew Cleugh, a boy ten or twelve years old; fifty-one to John Corry. Are now in same state as when brought to him; were perfect when taken from him—taken by the men on Saturday morning.

The lamps were bruised and bent—too much so ; and holes torn in the gauze. One had the oil-plug out; one a bend. One had signs of fire on the lower half of the gauze, as if gas had been burning against it; and there was also a round mark of the same kind on the side of the gauze, near the top, showing that that part had been over the top of the flame, and this mark corresponded with the crush the gauze and lamp had received. I believe that the lower oxidation shows that gas was in the mine. Two of the lamps were oily, as if they had lain on their side and the oil had flowed on to the gauze. The gauzes were good, and also the lamps, and there is every appearance of all having been in a good state before the accident.

Thursday, 10th.—Mr. Lyell and I went down the Haswell little pit, and carefully examined the workings; went down about eleven o'clock, and came up about

half-past six or seven o'clock. Whilst in the mine
heard one fall; and near Williamson judd were in
some danger from a fall that fell in the midst of us,
cutting off Mr. Lyell and some from myself and others.

Friday, 11*th*.—Again at the inquest, which was
resumed this morning.

George Hunter.—Colliery viewer for twenty-five
years, known coal mines for forty years. Examined
the pit on Tuesday with Mr. Wood. Agrees generally
with Mr. Wood, but always expects gas in the *goaf*,
if there be any gas at all; thinks the accident arose
from gas in the goaf, and a lamp injured by a fall.
When the barometer is low, gas appears. There are constant changes in the gas with high and low barometer;
when it has been high and a sudden fall comes, gas
appears. Men can light a pipe by a Davy lamp.
Smoking is strictly forbidden, but has known cases of
men smoking; men will smoke sometimes—is a very
great evil.

Jury would call no more witnesses, and gave a
verdict of *accidental death*. Fully agree with them.

Saturday, 12*th*.—Returned to London.

Five days' inquest. For reports, see 'Times,' October
2, 3, 4, 11, 12, 1844.

The following account was lately written by Sir
Charles Lyell at the request of a friend:—

'Faraday undertook the charge with much reluctance, but no sooner had he accepted it than he seemed
to be quite at home in his new vocation. He was seated
near the coroner, and cross-examined the witnesses
with as much talent, skill, and self-possession as if he
had been an old practitioner at the bar. We spent
eight hours,- not without danger, in exploring the

galleries where the chief loss of life had been incurred. Among other questions, Faraday asked in what way they measured the rate at which the current of air flowed in the mine.   An inspector took a small pinch of gunpowder out of a box, as he might have taken a pinch of snuff, and allowed it to fall gradually through the flame of a candle which he held in the other hand. His companion, with a watch, marked the time the smoke took going a certain distance.   Faraday admitted that this plan was sufficiently accurate for their purpose ; but, observing the somewhat careless manner in which they handled their powder, he asked where they kept it.   They said they kept it in a bag, the neck of which was tied up tight.   " But where," said he, "do you keep the bag ? "   " You are sitting on it," was the reply ; for they had given this soft and yielding seat, as the most comfortable one at hand, to the commissioner. He sprang up on his feet, and, in a most animated and expressive style, expostulated with them for their carelessness, which, as he said, was especially discreditable to those who should be setting an example of vigilance and caution to others who were hourly exposed to the danger of explosions. . . . . Hearing that a subscription had been opened for the widows and orphans of the men who had perished by the explosion, I found, on inquiry, that Faraday had already contributed largely. On speaking to him on the subject, he apologised for having done so without mentioning it to me, saying that he did not wish me to feel myself called upon to subscribe because he had done so.'

II.

His reputation was marked this year by his election as one of the eight foreign Associates of the Academy of Sciences, Paris. In answer to Mr. Magrath, who sent him from the ' Journal des Débats ' notice of his election, he said,—

' I received by this morning's post notice of the event in a letter from Dumas, who wrote from the Academy at the moment of the deciding the ballot, and, to make it more pleasant, Arago directed it on the outside.'

He was also made Honorary Member of the Sheffield Scientific Society.

A letter from Baron Humboldt, and one from Professor Liebig, are both characteristic of the writers; and one from Miss Edgeworth, is of some interest.

ALEXANDER HUMBOLDT TO FARADAY.

' Sans-Souci, le 12 mars 1844.

' Je sais que vous avez conservé, monsieur, beaucoup de bienveillance pour ma personne et mes travaux ; je mérite cette insigne faveur par l'admiration que je professe pour vous, parce que, un des premiers sur le continent, j'ai deviné combien votre nom deviendrait grand.

' J'écris ces lignes pour vous donner un petit signe de vie, et pour vous prier de recevoir avec bienveillance un de mes plus spirituels amis le docteur Carus (de Dresde), premier médecin du roi de Saxe, célèbre parmi nous par de beaux travaux de physiologie et de la plus fine anatomie des animaux d'un ordre inférieur.

'Permettez qu'il soit auprès de vous l'interprète des sentiments de respectueuse admiration que je vous ai voués pour la vie.

'Al. Humboldt.'

PROFESSOR LIEBIG (Æt. 41) TO FARADAY.

'Giessen: December 19, 1844.

'Dear Faraday,—

. . . . . .

'Nature has bestowed on you a wonderfully active mind, which takes a lively share in everything that relates to science. Many years ago your works imparted to me the highest regard for you, which has continually increased as I grew up in years and ripened in judgment; and now that I have had the pleasure of making your personal acquaintance, and seeing that in your character as a man you stand as high as you do in science, a feeling of the greatest affection and esteem has been added to my admiration. You may hence conceive how grateful I am for the proof of friendship which you have given me.

'I have every reason to be satisfied with my journey in Great Britain; rare proofs of recognition have indeed been given me. What struck me most in England was the perception that only those works which have a practical tendency awake attention and command respect; while the purely scientific, which possess far greater merit, are almost unknown. And yet the latter are the proper and true source from which the others flow. Practice alone can never lead to the discovery of a truth or a principle. In Germany it

is quite the contrary.   Here, in the eyes of scientific
men, no value, or at least but a trifling one, is placed
on the practical results.   The enrichment of science is
alone considered worthy of attention.   I do not mean
to say that this is better; for both nations the *golden
medium* would certainly be a real good fortune.   The
meeting at York, which was very interesting to me
from the acquaintance of so many celebrated men, did
not satisfy me in a scientific point of view.   It was
properly a feast given to the geologists, the other
sciences serving only to decorate the table.   The
direction, too, taken by the geologists appeared to me
singular, for in most of them, even the greatest, I found
only an empirical knowledge of stones and rocks, of
some petrifactions and few plants, but no science.
Without a thorough knowledge of physics and chemistry, even without mineralogy, a man may be a
great geologist in England.   I saw a great value laid
on the presence of petrifactions and plants in fossils,
whilst they either do not know or consider at all the
chemical elements of the fossils, those very elements
which made them what they are.

'Farewell, dear Faraday, preserve to me your
friendly favour, and believe me, with all sincerity, to
be, yours very truly,

'DR. JUST. LIEBIG.'

MISS EDGEWORTH TO FARADAY.

'Edgeworth Town: May 6, 1844.

'Dear Sir,—I am much gratified by your desire to
have my father's memoirs *as a souvenir from myself*,
and you shall have the assurance of my grateful regard

and high esteem under my own hand—a hand which never was put to a false compliment or an insincere profession.

'Were I writing to anyone but yourself, I would express without restraint or reserve, and with the warmth with which I feel it, admiration for talents and inventive genius directed to the best purposes, free from the petty envy and jealousy which too often cloud the lustre of genius and poison the happiness of the possessor.

'The brightness of your day, the cheerfulness of your temper even under the trials of ill-health, and the evident enjoyment you have in science and literature for their own sake, together with your love for your private friends and the serenity of your domestic life, prove (whatever Rousseau may have said or felt to the contrary) that "*Sois grand homme et sois malheureux*," is not the inevitable doom of genius.

.   .   .   .   .   .

'I am, dear Sir, sincerely yours,
                              'MARIA EDGEWORTH.'

A letter to his friend, Professor De la Rive, on the condensation of gases, belongs to this year ; and one to a noble lady of the highest talent, 'who proposed to become his disciple, and to go through with him all his own experiments.' These are pictures of his nature.

FARADAY TO PROFESSOR DE LA RIVE.

'Royal Institution : February 20, 1845.

'My dear de la Rive,—The thought of writing to you has been so constantly on my mind, and therefore by comparison so fresh, that I had no idea, until this

minute that I have looked at your letter, that I had
received it so long ago. I have waited and waited
for a result, intending to write off to you on the instant,
and hoping by that to give a little value to my letter,
until now, when the time being gone, and the result
not having arrived, I am in a worse condition than
ever, and the only value my letter can have will be
in the kindness with which you will receive it. The
result I hoped for was the condensation of *oxygen*, but
though I have squeezed him with a pressure of sixty
atmospheres, at the temperature of 140° F. below 0°,
he would not settle down into the liquid or solid state.
And now being tired and ill, and obliged to prepare
for lectures, I must put the subject aside for a little
while. Other results of this kind, i.e. of the liquefaction
and solidification of bodies usually gaseous, which I
have obtained, you will have seen noticed in the
" Annales de Chimie." The full account I hope to
send you soon from the " Philosophical Transactions."

' As to the ozone subject. It is exceedingly curious,
and I am really surprised to think how many results
and reasons there appear to be, all tending in one
direction, and yet without any one of them furnishing
an overruling and undeniable proof. I get confused
with the numerous reasons; my bad memory will not
hold them, and with my judgment longs to rest on
some one proof, such as a little ozone in the separate
visible or tangible state.

' Nitrogen is certainly a strange body ; it encourages
every sort of guess about its nature, and will satisfy
none. I have been trying to look at it in the condensed state, but as yet it escapes me.

' Your kind invitation for the scientific meeting in

August is very pleasant to the thought, but I dare not hope much for such happiness. I long to see Geneva and Switzerland again, but there are many things which come between me and my desires in that respect. I know the kindness of your heart, and how far I may draw upon you if I come; and I thank you most truly for not only the invitation you have sent me, but for all the favour you would willingly show me. Do you remember one hot day, I cannot tell how many years ago, when I was hot and thirsty in Geneva, and you took me to your house in the town, and gave me a glass of water and raspberry vinegar? That glass of drink is refreshing to me still.

'Adieu, my dear friend. Remember me kindly to Madame de la Rive; and, if I am not too far wrong in the collection of thoughts and remembrance of past things, bring me to the mind of one or two young friends who showed me a doll's house once, and with whom I played on the green.

'Yours most truly and affectionately,

'M. FARADAY.'

To a lady of the highest talent who proposed ' to become his disciple and to go through with him all his own experiments,' he wrote :—

'Royal Institution : October 24, 1844.

'Dear Lady——,—Your letter ought to have been answered before, but there are two circumstances which have caused delay—its high character and my want of health; for since I returned from a very forced journey to Durham, I have been under the doctor's hand. I am quickly recovering, and now have the difficult

pleasure of writing to you. I need not say how much
I value your letter—you can feel that; and even if it
were possible that you did not, no words of mine would
convey the consciousness to you : the thanks which I
owe you can only properly be acknowledged by an open
and sincere reply, and the absence of all conventional
phrase. I wonder that, with your high object, and
with views, determinations, and hopes consistent with
it—all of which are justified by the mind and powers
which you possess, which latter are not known to yourself only, but, as I say in perfect simplicity, are now
made fully manifest to others. I wonder that you
should think as I believe you do of me. But whilst
I wonder, and at the same time feel fully conscious of
my true position amongst those that think and know
how unworthy I am of such estimation, I still receive it
with gratitude from you, as much for the deep kindness
as for that proportion of the praise which I may perhaps think myself entitled to, and which is the more
valuable because of the worthiness of the giver.

' That with your deep devotion to your object you
will attain it, I do not doubt. Not that I think your
aspirations will not grow with your increasing state
of knowledge, and even faster than it ; but you must be
continually passing from the known to the unknown,
and the brightness of that which will become known,
as compared to the dulness, or rather obscurity, which
now surrounds it, will be, and is worthy to be, your
expected reward. And, though I may not live to see
you attain even what your mind now desires, yet
it will be a continually recurring thought in my
imaginings, that if you have life given you you will
do so.

' That I should rejoice to aid you in your purpose you cannot doubt, but nature is against you. You have all the confidence of unbaulked health and youth both in body and mind; I am a labourer of many years' standing, made daily to feel my wearing out. You, with increasing acquisition of knowledge, enlarge your views and intentions; I, though I may gain from day to day some little maturity of thought, feel the decay of powers, and am constrained to a continual process of lessening my intentions and contracting my pursuits. Many a fair discovery stands before me in thought which I once intended, and even now desire, to work out; but I lose all hope respecting them when I turn my thoughts to that one which is in hand, and see how slowly, for want of time and physical power, it advances, and how likely it is to be not only a barrier between me and the many beyond in intellectual view, but even the last upon the list of those practically wrought out. Understand me in this: I am not saying that my mind is wearing out, but those physico-mental faculties by which the mind and body are kept in conjunction and work together, and especially the memory, fail me, and hence a limitation of all that I was once able to perform into a much smaller extent than heretofore. It is this which has had a great effect in moulding portions of my later life; has tended to withdraw me from the communion and pursuits of men of science, my contemporaries; has lessened the number of points of investigation (that might at some time have become discoveries) which I now pursue, and which, in conjunction with its effects, makes me say, most unwillingly, that I dare not undertake what you propose—to go with you through even my own experiments. You do not know, and should not now but that I have no concealment on this point from you, how often I have to go to my medical friend to speak of giddiness and aching of the head, &c., and how often he has to bid me cease from restless thoughts and mental occupation and retire to the sea-side to inaction.

' If I were with you, I could talk for hours of your letter and its contents, though it would do my head no good, for it is a most fertile source of thoughts to my mind ; and whether we might differ upon this or that point or not, I am sure we should not disagree. I should be glad to think that high mental powers insured something like a high moral sense, but have often been grieved to see the contrary, as also, on the other hand, my spirit has been cheered by observing in some lowly and uninstructed creature such a healthful and honourable and dignified mind as made one in love with human nature. When that which is good mentally and morally meet in one being, that that being is more fitted to work out and manifest the glory of God in the creation, I fully admit.

' You speak of religion, and here you will be sadly disappointed in me. You will perhaps remember that I guessed, and not very far aside, your tendency in this respect. Your confidence in me claims in return mine to you, which indeed I have no hesitation to give on fitting occasions, but these I think are very few, for in my mind religious conversation is generally in vain. There is no philosophy in my religion. I am of a very small and despised sect of Christians, known, if known at all, as *Sandemanians*, and our hope is founded on the faith that is in Christ. But though the natural works of God can never by any possibility come in contradiction with the higher things that belong to our future existence, and must with everything concerning Him ever glorify Him, still I do not think it at all necessary to tie the study of the natural sciences and religion together, and, in my intercourse with my fellow creatures, that which is religious and that which is philosophical have ever been two distinct things.

'And now, my dear Lady, I must conclude until I see you in town ; being *indeed* your true and faithful servant,

'M. FARADAY.'

CHAPTER III.

LATER PERIOD OF ELECTRICAL RESEARCH—DISCOVERY OF THE
'MAGNETISATION OF LIGHT'—THE MAGNETIC STATE OF ALL
MATTER—ATMOSPHERIC MAGNETISM.

THE second period of Faraday's electrical work lasted ten years. The discoveries he made were published in the 'Philosophical Transactions.' They constitute from the nineteenth to the thirtieth series of his 'Experimental Researches in Electricity.' The three great results which he obtained he called ' the magnetisation of light,' 'the magnetic condition of all matter,' and ' atmospheric magnetism.'

1845.
ÆT.53–54.

Faraday's reputation at this time was so great that it added to the renown which followed the publication of each of these new discoveries; but great as the results were, they will not at the present time rank with the three great discoveries of ' magneto-electricity,' ' voltaic induction,' and ' definite electro-chemical decomposition,' which made the glory of the first period of the ' Researches in Electricity.'

I.

In the beginning of 1845 Faraday worked on the condensation of gases; on August 30 he began to experiment on polarised light and electrolytes, a subject which in 1833 had given ' no result.' After three days he worked with common electricity, trying glass, heavy optical glass, quartz, Iceland spa. Still he got no effect on the polarised ray.   On September 13 he writes : ' To-day worked with lines of magnetic force, passing them across different bodies transparent in different directions, and at the same time passing a polarised ray of light through them, and afterwards examining the ray by a Nichol's eye-piece or other means.   Air, flint-glass, rock-crystal, calcareous spa, were examined, but without effect.

' Heavy glass was experimented with.   It gave no effects when the *same magnetic poles* or the *contrary* poles were on opposite sides (as respects the course of the polarised ray), nor when the same poles were on the same side either with the constant or intermitting current ; **BUT** when contrary magnetic poles were on the same side there *was an effect produced on the polarised ray*, and thus magnetic force and light were proved to have relations to each other.   This fact will most likely prove exceedingly fertile, and of great value in the investigation of conditions of natural force.'   He immediately goes on to examine other substances, but with ' no effect ;' and he ends saying : ' Have got enough for to-day.'

On September 18 he makes out more closely the circumstances and laws of action, and ' does *an excellent day's work* ; ' and then for four days he works out his results.   On September 30 he writes : ' So the combination of electric currents with magnetic forces does not give any very striking effects, and perhaps there are none which polarised light can show.   But I am not sure of that.   The quantity of fluid was very small for length of ray to pass through (for in making it more I should have weakened the magnetic curves); and considering the nature of the relation between magnetic

and electric forces, I think there must be some effect
produced which stronger magnets, and other forms of
apparatus, and the progress of our knowledge, will
enable us hereafter to develope.'

' Still I have at last succeeded in illuminating a
magnetic curve or line of force, and in magnetising
a ray of light.'

' What effect does this force have in the earth where
the magnetic curves of the earth traverse its substance ?
also what effect in a magnet ? '

' Does this force tend to make iron and oxide of iron
transparent ? '

During October, for six days, he worked at this subject ; and on November 6 he sent the nineteenth series of
' Researches in Electricity' to the Royal Society, on the
magnetisation of light and the illumination of the lines
of magnetic force.

It begins thus :

' I have long held an opinion, almost amounting to
conviction, in common I believe with many other lovers
of natural knowledge, that the various forms under which
the forces of matter are made manifest have one common origin ; or, in other words, are so directly related
and mutually dependent that they are convertible, as it
were, one into another, and possess equivalents of power
in their action.

' This strong persuasion extended to the powers of
light, and led to many exertions having for their object
the discovery of the direct relation of light and electricity.   These ineffectual exertions could not remove my
strong persuasion, and I have at last succeeded.

' Not only heavy glass, but solids and liquids, acids
and alkalies, oils, water, alcohol, ether, all possess this

power. I have not been able to detect the exercise of
this power in any one of the substances in the gaseous
class.'

Then he showed that all bodies were affected by
helices as by magnets. 'The causes of the action are
identical as well as the effects.'

In conclusion he says : 'Another form of the great
power is distinctly and directly related to the other
forms ; or the great power manifested by particular
phenomena in particular forms is here further identi-
fied and recognised by the direct relation of its form
of light to its forms of electricity and magnetism.'

On November 3, a new horse-shoe magnet came
home, and Faraday immediately began to experiment
on the action of the polarized ray on gases, but with no
effect.

The following day he repeated an experiment which
had given no result on October 6. A bar of heavy
glass was suspended by silk between the poles of the
new magnet. ' When it was arranged, and had come
to rest, I found I *could* affect it by the magnetic forces
and give it position. Thus touching diamagnetics by
magnetic curves, and observing a property quite independent of light, by which also we may probably trace
these forces into opaque and other bodies, as the metals,
&c.' Then he describes how the heavy glass was repelled
from the poles.

On November 7 he takes up his new discovery.
' First of all, the great fact of the 4th was verified ; then
borate of lead, rock-crystal, flint-glass, sulphur, india-
rubber, sulphate of lime, asbestus, jet, all acted as heavy
glass. If a man could be in the magnetic field, like
Mahomet's coffin, he would turn until across the

magnetic line.' All kinds of substances and metals
proved to be like the heavy glass.

Dr. Tyndall says :—

' And now theoretic questions rush in upon him. Is this new force a true repulsion, or is it merely a differential attraction ? Might not the apparent repulsion of diamagnetic bodies be really due to the greater attraction of the medium by which they are surrounded ? He tries the rarefaction of air, but finds the effect insensible. He is averse to ascribing a capacity of attraction to space, or to any hypothetical medium supposed to fill space. He therefore inclines, but still with caution, to the opinion that the action of a magnet upon bismuth is a true and absolute repulsion, and not merely the result of differential attraction. And then he clearly states a theoretic view sufficient to account for the phenomena. " Theoretically," he says, " an explanation of the movements of the diamagnetic bodies, and all the dynamic phenomena consequent upon the action of magnets upon them, might be offered in the supposition that magnetic induction caused in them a contrary state to that which it produced in ordinary matter." That is to say, while in ordinary magnetic influence the exciting pole excites adjacent to itself the contrary magnetism, in diamagnetic bodies the adjacent magnetism is the same as that of the exciting pole. This theory of reversed polarity, however, does not appear to have ever laid deep hold of Faraday's mind ; and his own experiments failed to give any evidence of its truth. He therefore subsequently abandoned it, and maintained the *non-polarity* of the diamagnetic force.'

In Faraday's notes on November 19, he says: 'Hence the power of the surrounding medium to cause the vertical attraction or repulsion of other bodies is very manifest, and as yet everything seems to show that these actions are of exactly the same nature as those with bismuth, &c., in air. It would be very curious if, after iron, nickel and cobalt, the air were to prove the next most magnetic body ; but perhaps some of the metals, as platinum or silver, may come in.' And in a few lines he adds : ' Air is becoming very important. Oxygen and nitrogen may have great differences between them, the magnetic power of one neutralising the deficiency of power in the other. As it is, air is the most magnetic of all earthly bodies except iron, nickel and cobalt. Not unlikely that the *earth's magnetism* may reside essentially in *the air*. All the gases require careful examination and consideration. The magnetic condition and relation of the air, gases, and vapours, probably a very fine separate subject.'

On December 6 he sent the twentieth series and on December 24 the twenty-first series of ' Experimental Researches ' to the Royal Society, on new magnetic actions and on the magnetic condition of all matter.

In December he sent the following sealed packet to Sir John Herschel, but at the same time he said that Herschel might open it if he pleased.

'December 22, 1845.

' I have reason from experiment to think that a ray is not indifferent as to its line of path, but has different properties in its two directions, and that by opposing rays endways, new results will be obtained. I have ordered apparatus already for the experimental investigation of this point, and only want *time*.

'I have already made a certain progress in the   1845.
endeavour to obtain electric currents or magnetic force   Æт.53–54.
from light, by the use of circular polarisation natural and
constrained, and also on other principles which I need
not advert to here.

'M. FARADAY.'

SIR JOHN HERSCHEL TO FARADAY.

'Collingwood: January 22, 1846.

' My dear Sir,—You did not surely think me so in-
curious, or rather so deficient in interest, respecting the
astonishing series of discoveries into which you are now
entered fairly, that, having your express permission to
open and read the paper you sent me sealed, I should
not avail myself of it.   Accordingly, I have done so,
but I thought it best to reinclose it to you, rather than
consign it to the flames, which I would not do to a bit
of your hand-writing.   Should the first of your views
expressed in it be really verified, a new field of speculation on the nature of light will be opened, as I do
not understand what the undulatory or indeed any
theory can have to say to a fact of that nature.

' Go on and prosper—"from strength to strength,"
like a victor marching with assured step to further con-
quests; and be assured that no voice will join more
heartily in the peans that already begin to rise, and
will speedily swell into a shout of triumph, astounding
even to yourself, than that of

' Yours most truly,

' J. F. W. HERSCHEL.'

He this year published a letter to M. Dumas on the
liquefaction of gases, in the 'Annales de Chimie;' and

papers on the ventilation of the coal-mine goaf, and on the magnetic relations and characters of the metals, in the ' Philosophical Magazine ; ' and additional remarks respecting the condensation of gases in the ' Proceedings' of the Royal Society.

His first Friday discourse was on Haswell Mine accident. In this he said : ' Coal mines are out of society's thoroughfare and criticism, and so improve slowly only.' His second discourse was on condensed gases. His third was on anastatic printing—' Problem stated as the transfer of a printed page.' The last, on the Artesian well and water in Trafalgar Square.

He gave the Christmas Lectures on chemistry.

For the Trinity House he only made a long and exact comparison of the consumption and light of sperm and rape oil.

II.

The following letters from Sir John Herschel, Dr. Whewell, and Mrs. Marcet, may be looked at as marks of his reputation.

SIR JOHN HERSCHEL TO FARADAY.

'Collingwood, Hawkhurst, Kent : November 9, 1845.

' My dear Sir,—I have this morning read with great delight a notice in the " Athenæum " of your experiments proving the connection of light with magnetism. In the first place, let me congratulate you cordially on a discovery of such moment, which throws wide a portal into the most recondite arcana of nature. If I understand rightly the very meagre account given of your discovery, it amounts to this—that the electro-magnetic current is capable of causing *the plane of polarisation* of a ray of light to revolve, for I can find no other probable interpretation of the expression " a beam of polarised light is deflected by the electric current, so that it may be made to revolve between the poles of a magnet."

' If this be really the state of the case, it is what I have long anticipated as extremely likely—indeed, almost certain—to be sooner or later experimentally demonstrated. *Voici mes raisons.*

' There are three distinct classes of phenomena in which a helicoidal dissymmetry occurs. 1st, the plagie-dral faces on crystals, such as quartz, which belong to an otherwise symmetrical system. These faces in some crystals indicate a right-handed, in others a left-handed dissymmetry of the helicoidal kind. 2ndly, the rotation of the plane of polarisation of a ray of light, when transmitted through certain solids and liquids, in-dicating a helicoidal dissymmetry both in the ray and in the molecules, or at least a capacity in the ray to be affected by that peculiarity in the latter. 3rdly, in a rectilinear electric current, which, deflecting a needle in a given direction, as to right and left, all around it, indicates again a dissymmetry of the same kind.

' Now, I reasoned thus :—Here are three phenomena agreeing in *a very strange peculiarity*. Probably, this peculiarity is a connecting link, physically speaking, among them. Now, in the case of the crystals and the light, this probability has been turned into cer-tainty by my own experiments. Therefore, induction led me to conclude that a similar connection exists, and must turn up somehow or other, between the electric current and polarised light, and that the plane of polarisation would be deflected by magneto-electricity.

'It is now a great many years ago that I tried to bring this to the test of experiment (I think it was

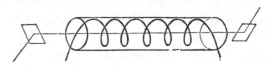

between 1822 and 1825), when, on the occasion of a great magnetic display by Mr. Pepys, at the London Institution, I came prepared with a copper helix in an earthern tube (as a non-conductor), and a pair of black glass plates, so arranged as that the second reflection should extinguish a ray polarised by the first. After traversing the axis of the copper helix, I expected to see light take the place of darkness—perhaps coloured bands—when contact was made. The effect was *nil*. But the battery was exhausted, and the wire long and not thick, and it was doubtful whether the full charge remaining in the battery *did* pass, being only a single couple of large plates.

'There remained to be made another experiment before a negative could be considered as proved—viz., to make the light move along a straight wire or a combination of such : thus—

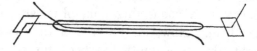

but this requiring preparation on the spot, I could not then make, and have never since had an opportunity ; but the subject has often recurred to my mind, and I have made frequent mention of it, as a line of experiment worthy to be pursued.

' You will be disposed to ask to what all this tends. Assuredly not to interfere for a moment with your claim to a beautiful discovery (for though I may regret that I did not prosecute a train of inquiry (which seemed so promising) up to a decisive fact, I consider it honour enough to have entertained a conception which your researches have converted into a reality) ; but if it be not presumptuous in me to suggest a line of inquiry to you, I would willingly draw your attention to the other member of the triple coincidence above alluded to.

' *There can be now no doubt* of the connection of the *crystalline forces* with magnetism and electricity. It seems to me now all but certain that the space or ether (?) surrounding an electrified wire or a magnet is in the same state with the space or ether intervening between the molecules of a plagiedral crystal. Polarised light *is the test of that state*—a helicoidal, dissymmetrical state. This is the mode in which the phenomena present themselves to my mind—not that light *is* electricity or magnetism, but that it is affected by them *as* by certain forms of matter ; *which*, therefore, I conclude to be under the influence of magnetic currents, in some concealed way circulating about them ; and the line of inquiry I allude to, is to ascertain whether the crystals formed under the direct influence of magnetic currents or between the poles of magnets may not be thereby made to assume plagiedral faces or show other indications of a symmetrical action. If so, the existence of the plagiedral faces on quartz is accounted for by the presence of such currents during their formation.

'Believe me, my dear Sir, yours most truly,

'J. F. W. HERSCHEL.'

On the relation of light to magnetism, Dr. Whewell wrote to Faraday :—

'Trinity Lodge, Cambridge : November 20, 1845.

' My dear Sir,—I am somewhat scrupulous about trying to take up your time with letter writing, but I cannot help wishing to know a little more than the "Athenæum" tells us as to your recent discoveries of the relations of light and magnetism. I cannot help believing that it is another great stride up the ladder of generalisation, on which you have been climbing so high and standing so firm. I do not ask you to take the trouble of telling me what your discovery is, but perhaps you may be able to tell me where now, or in a short time, I may see some distinct account of it.

' I hope you will have health and strength granted you to follow out this and many more great discoveries.

'Believe me, my dear Sir, yours very truly,

' W. WHEWELL.'

MRS. MARCET TO FARADAY.

'Danesbury, Welwyn, Herts : November 24, 1845.

' Dear Mr. Faraday,—I have this morning read in the " Athenæum," some account of a discovery you announce to the public respecting the identity of the imponderable agents, heat, light, and electricity ; and as I am at this moment correcting the sheets of my " Conversations on Chemistry " for a new edition, might I take the liberty of begging you would inform me

where I could obtain a correct account of this dis- 1845.
covery? It is, I fear, of too abstruse a nature to be Æt. 54.
adapted to my young pupils; yet I cannot make up
my mind to publish a new edition without making
mention of it; I have, therefore, kept back the proof
sheets of the "Conversation on Electricity," which I was
this morning revising, until I receive your answer, in
hopes of being able to introduce it in that sheet.

‘ Believe me, dear Mr. Faraday, very truly yours,
‘ JANE MARCET.’

The ‘Athenæum,’ November 8, 1845, said :—

‘ Mr. Faraday, on Monday (November 3), announced
at a meeting of the council of the Royal Institution a
very remarkable discovery, which appears to connect
the imponderable agencies yet closer together, if it
does not indeed prove that light, heat, and electricity are merely modifications of one great universal
principle.’

The minutes of the general monthly meeting R.I.
November 3, 1845, are these :—

‘ Mr. Faraday communicated to the meeting the results of some recent researches on the correlation of
magnetism and light.’

This year he was made Corresponding Member of
the National Institute, Washington ; and of the Société
d'Encouragement, Paris.

III.

He shows something of his nature at this time in a
few letters, and in a journal which he kept during a
tour in France. He went partly to inspect the French

lighthouses, partly to be admitted into the French Academy. He was away three weeks, with Mrs. Faraday and Mr. G. Barnard.

A glimpse of his affection to his early friend Huxtable here once more occurs :—

FARADAY TO HUXTABLE.

'Royal Institution : June 11, 1845.

' Dear Huxtable,—I intended to have seen you before this, but am somewhat held at home by an invalid friend, so that I have not been able. But I cannot longer refrain from sympathising deeply with your grief in your loss. I heard of it on Friday evening, and it came on me very suddenly, for I had not thought of such an end, at least as yet. But in life we are in death, and these things ought never to be altogether away from our thoughts. If occasion serves (otherwise do not trouble yourself), express my sorrow to the members of your family, for though circumstances have made it long since I have seen them, I know that all were and are bound together by the most affectionate ties, and that they will greatly feel this deprivation.

' Ever, dear Huxtable, yours truly,

'M. FARADAY.'

FARADAY TO PROFESSOR AUG. DE LA RIVE.

'Brighton : December 4, 1845.

' My dear Friend,—Your letter, which I received this morning, was a very great gratification to me, not more for the approbation which it conveyed than for the kindness with which I know it is accompanied. I count upon you as one of those whose free hearts have pleasure in my success, and I am very grateful to you for it. I have had your last letter by me on my desk for several weeks, intending to answer it; but absolutely I have not been able, for of late I have shut myself up in my laboratory and wrought, to the exclusion of everything else. I heard afterwards that even your brother had called on one of these days and been excluded.

1845.
Æt. 54.

' Well, a part of this result is that which you have heard, and my paper was read to the Royal Society, I believe, last Thursday, for I was not there ; and I also understand there have been notices in the " Athenæum," but I have not had time to see them, and I do not know how they are done. However, I can refer you to the " Times " of last Saturday (November 29th) for a very good abstract of the paper. I do not know who put it in, but it is well done though brief. To that account, therefore, I will refer you.

' For I am still so involved in discovery that I have hardly time for my meals, and am here at Brighton both to refresh and work my head at once, and I feel that unless I had been here, and been careful, I could not have continued my labours. The consequence has been that last Monday I announced to our members at the Royal Institution another discovery, of which I will give you the pith in a few words. The paper will go to the Royal Society next week, and probably be read as shortly after as they can there find it convenient.

' Many years ago I worked upon optical glass, and made a vitreous compound of silica, boracic acid, and lead, which I will now call heavy glass, and which Amici uses in some of his microscopes ; and it was this substance which enabled me first to act on light by magnetic and electric forces. Now if a square bar of this substance, about half an inch thick and two inches long, be very freely suspended between the poles of a powerful horse-shoe electro-magnet, immediately that the magnetic force is developed, the bar points; but it does not point from pole to pole, but equatorially or across the magnetic lines of force, i.e. east and west in respect of the north and south poles. If it be moved from this position it returns to it, and this continues as long as the magnetic force is in action. This effect is the result of a still simpler action of the magnet on the bar than what appears by the experiment, and which may be obtained at a single magnetic pole. For if a cubical or rounded piece of the glass be suspended by a fine thread six or eight feet long, and allowed to hang very near a strong magneto-electric pole (not as yet made active), then on rendering the pole magnetic, the glass will be repelled and continue repelled until the magnetism ceases. This effect or power I have worked out through a great number of its forms and strange consequences, and they will occupy two series of the "Experimental Researches." It belongs to *all matter* (not magnetic, as iron), without exception, so that every substance belongs to the one or the other class—magnetic or diamagnetic bodies. The law of action in its simple form is that such matter tends to go from strong to weak points of magnetic force, and in doing this the substance will go in either direction along the magnetic curves or in either direction across them. It is curious that amongst the metals are found bodies possessing this property in as high a degree as perhaps any other substance. In fact, I do not know at present whether heavy glass, or bismuth, or phosphorus is the most striking in this respect. I have very little doubt that you have an electro-magnet strong enough to enable you to verify the chief facts of pointing equatorially and repulsion, if you will use bismuth carefully examined as to its freedom from magnetism, and making of it a bar an inch and a half long, and one-third or one-fourth of an inch wide. Let me, however, ask the favour of your keeping this fact to yourself for two or three weeks, and preserving the date of this letter as a record. I ought (in order to preserve the respect due to the Royal Society) not to write a description to any one until the paper has been received or even read there. After three weeks or a month, I think you may use it, guarding, as I am sure you will do, my right. And now, my dear friend, I must conclude, and hasten to work again. But first give my kindest respects to Madame de la Rive, and many thanks to your brother for his call.

' Ever your obedient and affectionate friend,

' M. FARADAY.'

To Professor Wheatstone, who told him of Becquerel's researches on the magnetic condition of all matter, Faraday writes :—

FARADAY TO C. WHEATSTONE, ESQ.

' Royal Institution : Friday night, December 5, 1845.

' Many thanks, my dear Wheatstone, for your note. I have in consequence seen Becquerel's paper, and added a note at the first opening of my paper. It is astonishing to think how he could have been so near the discovery of the great principle and fact, and yet so entirely miss them both, and fall back into old and preconceived notions.

' Ever truly yours,

' M. FARADAY.

A few extracts from his French Journal will show what he was at this time.

Monday, July 7th.—Left London by Folkestone train at eight o'clock.

July 10th.—Fécamp.   On the pier is a small harbour light, fourth-class of Fresnel, to indicate the height of the tide.   On the height above is the chief lighthouse, first-class Fresnel.   It is a stone tower with the staircase inside, and the dwellings of the keepers at the bottom arranged about it.

Then he gives a full description of it and its arrangement.

After leaving the lighthouse, which we did in very bad, rainy weather, we continued our way in the cabriolet towards Etritat, over a road which so jolted and tossed us about that not only did it occasion many strange similes and figures of speech, but stirred up my bad memory, so that I think I shall not easily forget it.

Etritat.   This is a very curious place.   Whilst waiting for our meal, there was much bustle amongst the women, and on stepping out from between a number of boats thatched over on the shore, we found that the fishing boats were in the act of arriving, and were being drawn up a steep shingle beach by cables and capstans on shore, the latter being most actively manned by women. They seemed thoroughly well to understand what they were about, and the boats were soon hauled up; as many as six capstans, each with sixteen or twenty persons, chiefly women, going at once.   Then came the cargo, chiefly mackerel, and very fine; there were the douane officers moving about to see that no smuggling was in hand.   How the fish was disposed of I cannot

tell, but in a very short time it was in the yard of our
inn, and as many as fifteen or sixteen persons were
engaged in washing it and packing it in little baskets,
each holding from sixteen to twenty, curiously curved
in on purpose, and then neatly covered over and tied
up with very clean straw and string, in a peculiarly tech-
nical manner. Whether there were near to one hundred
or two hundred of these baskets I do not know, but
suddenly they were all carefully packed in a two-
wheeled cartwaggon with much straw. Four horses
were harnessed to the vehicle, the master of the inn
took his seat on the top, and off they set for Fécamp.
How they would bear the jolting that we had had
before them, I do not know. Two other changes of
horses, we were told, would be required in the course
of the night to carry them fast to Rouen. The object
was to get them there by six o'clock if possible, but
sometimes they cannot get there before seven o'clock.
This difference of an hour would make a difference of
200 francs in the value of the fish.

Walking out again, I found a number of women
washing on the sea shore. The river which runs down
the valley penetrates the shingle and chalky bank at
the valley's mouth and issues out from between the
shingles near low water mark. So at low water the
women came with their clothes to arrange for washing.
For this purpose they throw out the shingles where the
stream runs over a space two or three feet diameter,
casting the shingles seaward, and so quickly make a
basin, and this is filled with the purest soft water,
rapidly flowing in on the one side and out at the other.
Here they wash away, some rubbing with soap and
kneading the clothes on the side shingles, others thump-

ing the wet things with heavy wooden bats, and others rinsing out the washed things in the clear pools and then spreading them out higher up on the shingles to dry. The women are remarkably well and fittingly dressed, and neat and clean in their persons, especially about the feet. I saw no rags in wear. Their clothing is, externally, principally woollen, excellent in shape and fitness, and gay in colour. Their stockings quite. clean and neat, and their clogs quite black. It was really a most interesting and beautiful sight to see them either at the capstans or at the washing.

The rocks about here are chalky, very full of flint, very high, and the cliff presented extraordinary cases of natural arches and caverns, surpassing in size anything I recollect ever to have seen. The scale on which the strata break and divide is enormous. After a walk out towards the southern cliffs and amongst the lobster and crab deposits, I returned home, and we went to bed by daylight, i.e. by evening light. Weather bad, and has been very unsteady these last two days.

Friday, 11*th*.—George and I walked out before breakfast to the southern side of the little bay, and, it being low water, creeped round the point, and had a fine view of the natural pinnacles and arches beyond. They are wonderful in size. On our return, breakfasted very well, and then we all went out to see the women at their washing again. Most curious. The women here have their daily avocations governed by the tides; for they help to launch the boats and to haul them up —two very chief occupations for the day, and which govern most of their domestic duties. And then also their own peculiar occupation of washing is of necessity governed by the same great natural phenomenon. I asked the landlord what was the time of low water, and he went to ask his wife, saying she could tell.

Havre. All here looks like business. Yet even here the French do their work with a kind of gaiety which is very agreeable to look at, and seems better for the spirit than that determinate, and I may say desperate devotion to labour and profit, which marks London occupation, and I suppose still more American exertion. A man may make more money and even make his fortune quicker in the latter way, but it is very doubtful whether he is the happier man. What he gains in metal, I suspect he loses in spirit, and I think he loses on the whole.

Saturday, 12th.—Here are two large lighthouses at La Here, close to Havre, but at present works are in progress there and they are not allowed to be seen.

About twelve o'clock we left Havre by steamboat for Caen. As we started from the quay, one of the men on board took up a tambourine, and gave us, the ship, and the public the benefit of a kind of roll with certain fantastic variations. To be sure, it did not seem quite in keeping, that a good and powerful steamship, carrying a valuable cargo of cotton and passengers, should leave such a port as Havre to the sound of a tambourine ; neither was it at all in keeping with the heavy rolling and pitching which we found outside the harbour ; but still it was effect, though on a very small scale.

Monday, 14th.—This morning, George and I set off early by the diligence, leaving my wife at Caen in the hotel for three days, our object being to pass over the ground quickly to the lighthouses at Barfleur and Cap la Hève. We were in the rotonde or back part of the diligence by six o'clock, eight of us being closely packed up there. Of these, two were a newly married loving couple, and it was most amusing (though the amusement was rather strong in character) to see their enjoyment, especially that of the lady. She had by no means a sentimental form, or face, or manners, but such delight as she seemed to find in having a husband was most remarkable proof of her simplicity (perhaps). She talked to a third person, a young man, of how she liked to be loved. She showed off her husband's little watch and chain. She referred to his budding whiskers, his promise of a moustache, and *her* hopes of his having a beard, handling the parts, and referring to the places and forms she should like them to have, in a manner so illustrative and direct as to be worthy of a lecturer. The two fed each other with cherries and bread, and after a hearty meal at one place they comforted each other in the best way they could upon the occurrence of a little indigestion. The young man seemed rather ashamed now and then, and by an appearance of sleep gave some intimations, if not to his wife, at least to others; but it would not do, and he gave way very good-humouredly to his *bien aimé*. We breakfasted at St. Lo, a beautiful place ; and still further on, left the diligence at Valogne, and after a while proceeded on our way in a cabriolet.

The women's caps here are wonderful. We saw in the course of the day some more and more wonderful, but at Valogne came the finest and largest of all. It was on the top of a rich woman, as the hostess told us, who actually had her carriage there, and was moving about the town buying in things. As she came along the street, its height, and stiffness, and ex-

pansion, and the wind, actually brought her up.   She
had to clap her hands to her head and stand still for a
while, after which, by a series of tacks, she came up to
and into the little café where we were.   The cap was
handsome in materials and ornamented with coloured
ribbon, which is not usual.

During our ride (to Barfleur) we saw abundance of
that charming luminary, the glowworm.   The driver
called it the 'night worm.'

The lighthouse is situated on a rock, and is connected
by a granite causeway with the land, but in rough
weather and at high tide the sea makes a clean breach
over this and the rocks.   (Then he describes it fully.)

. . . .

Cap la Hève.   There at (Andeman) the land's end
and corner, we found rocky country, a curious little
fisherman's bay, a perfectly clear water, a tumbling sea,
rough men, and a lighthouse on a rock in the sea
about a mile out from shore, and looking beautifully in
the declining sunbeams.   A boat and three men took us
to the rock, on which, with some care and trouble, we
landed, and then the boat pushed off, to prevent her from
being knocked to pieces by the vigorous waves, for here
they leaped about in every direction.   (Then he describes
the lighthouse.)

We returned to Cherbourg by ten.

July 18*th*, *Friday*.—Honfleur, Rouen.   Entered a
steamboat at seven o'clock, and went up the Seine,
with a beautiful morning and beautiful scenery.   The
tide was up, so that all seemed perfection.   The water
appeared to bathe the feet of the trees.   The hills
are chalk, and on the banks of the river the cliffs

assume beautiful varieties of form, the green verdure of the sides intermingling delightfully with the white surface of fractures.

Arrived at Rouen, which appeared in its glory.

Saturday, 19*th*.—Have had a good night's rest, but was wakened up by a great talking in the street below, at an early hour, and found it was a company of sweepers about thirty in number, who were busy cleaning the streets, and talking together most vigorously. The continued talk and chatter of people together when they are at work, whether in the house, or in the street, or in the air, as upon some of the cathedral scaffolds, is quite remarkable. They must be a very thinking race to have so much to say.

July 22*nd*.—Paris. · Went to the Post Office, and the bonnet shop. It takes a good while to go about this large city. George and I wanted hats, and seeing a shop with a good number to choose from, which we had not seen before, went in and had begun a process of purchasing *chapeaux d'occasion* before we discovered what *occasion* meant. Finding they were *second-hand*, we were happy to be able to retreat as we entered.

Went to the Observatory, and found M. Arago, who was very kind and pleasant. I find that the meeting of the Academy of Sciences was yesterday, and that because of the July fêtes there will not be another until Wednesday (30th), which is very awkward for me, for I wanted to have left Paris on the Monday. Many things are deranged by the preparation for this fête.

Wednesday, 23*rd*.—Found Chevreul at the Jardin des Plantes, and had a long chat. Dumas not there. Returned, and found Dumas at our hotel. He dined with us, and afterwards I went with him to the Society of Encouragement, of which he took the chair, and I was made a member.

Thursday, 24*th.*—Bought some portraits. Went to Observatory to hear Arago give an astronomical lecture. He delivered it in an admirable manner to a crowded audience. Its object was to prove that the sun was the centre of the solar system, and to illustrate and prove the necessity of time in the passage of light, using for that purpose the phenomena of the eclipses of Jupiter's satellites. From thence went to Dumas' house at the Jardin des Plantes. Dined at Dumas' house. Dumas and I went to M. Milne Edwards.

Friday, 25*th.*—Went out and called on Biot, who appears to age, but was very cheerful.

Monday, 28*th.*—Went to Mr. Henry Laponte, and had a very pleasant conversation with him about the light-house apparatus. Arranged to go to the manufactory on Wednesday morning, for the fête to-morrow interferes with everything. Called on and saw Peltier. Called at the Conservatoire des Arts et Métiers, but Pouilliet and Payen not at home. Went to Dumas at the Jardin des Plantes ; found all at home ; they form a very pleasant family. · Went to the Observatory, and found Arago, who has called twice on us, and, moreover, has obtained tickets for us for special seats to see the fête on the ·Seine in the afternoon, and the fireworks at the Invalides in the evening. Called also on M. L. Fresnel (who has called on us), and, finding him at home, had a long talk with him. He is· to meet us at the shops of M. Laponte on Wednesday.

. . . . . .

At half past seven o'clock we went to the Sorbonne to see some very beautiful experiments on the application of the voltaic light to the solar microscope. The battery consisted of seventy-two pairs of Bunsen's plates.

Tuesday, 29*th*.—In the evening we all went to a station (having tickets for it from M. Arago) to see the illuminations on the river and the bridges, and also the fireworks. All went on well, and as daylight decreased the illuminations became beautiful. Boats floated about the river, all illuminated with lamps and lanthorns, and at last the fireworks began. Stars innumerable shot up from all parts of the river, and the effect gradually increased by the inflammation of pieces more and more large and powerful; concluding at last with such bursts of rockets, mines, &c., as I think equal those I saw at Rome on the occasion of the Pope's return to it. There appeared to be no accidents, and there was no uncomfortable crowding. Great precautions had been taken, for last year several deaths occurred from pressure in the narrow streets. Soldiers were about everywhere, both horse and foot. It is a curious feature to see the necessity for such precautions even in the midst of a holiday founded in commemoration of a successful revolution.

Wednesday, 30*th*.—Went to Mr. H. Laponte, at his workshops. He now both makes the optical and the mechanical part, and his workshops and apparatus are excellent, far surpassing anything we have in England. He had several of his lamps and apparatus in order. In the lamps colza oil is used, and they were all burning well, and, except one, without any sensible smoke. He was most kind in his attentions.

M. Fresnel did not arrive, and therefore I could

not have my difficulty about the expression of the power of the light solved, but we made some experiments, and I saw the mode, at all events.

Next we went to the well of Grenelle. The well is a wonderful work, penetrating 1,700 feet downwards into the earth, and rising 100 feet above it. We ascended the scaffold, and saw the water flowing up majestically through the pipe, and, after being measured, passing away down another pipe to the district of Paris about the Pantheon, which is the highest ground. And then to put one's hand in the water and feel it warm, and so to become really conscious that that clear black welling fluid, which is for ever silently flowing towards us, is bringing with it heat from the inner parts of the earth. It is very wonderful.

The view of Paris is very fine here.

From thence we went to the water-works at Chaillot, where an old steam engine of 80 horse-power draws water from the Seine and lifts it into reservoirs, from whence it runs to supply certain fountains in the gardens of the Tuileries and also certain parts of Paris. The engine has an enormous beam of wood, is very ancient in its appearance, and is, I believe, the first condensing engine which was made in France.

After this, went home for a while, and then to the Institute, to the sitting which, because of the fête yesterday, is held to-day instead of last Monday. Many of the members were gone out of town, but all that were there received me very kindly. I was glad to see Thénard, Dupuis, Flourens, Biot, Dumas of course, and Arago, Élie de Beaumont, Poinsot, Babinet, and a great many others whose names and faces sadly embarrassed my poor head and memory. Chatting together, Arago told me he was my senior, being born in 1786, and consequently 59 years of age.

Friday, August 1st.—Had a beautiful evening passage across the Channel to Folkestone. In the passage saw the lights at Cape Grinez on the French coast, and those at the South Foreland and also at Dungeness on the English land. In this part of the Channel the French lights revolve and the English lights are fixed. The light at Cape Grinez was very fine in its effect, and the flashes beautiful. It is one of the first order. Fresnel's construction. Of the two lights at the South Foreland the lower was in some positions the best, and in other positions it was the upper that excelled. The captain of the steamer thought that upon the whole the upper was the better light. The Dungeness light was feeble by comparison with the others, but this is in a great measure due to its low altitude and position.

Saturday, 2nd.—By two o'clock or soon after were at home in the Institution, where we felt **we** ought to be. We left George at the London Bridge station. Thanks be to him for all his kind care and attention on the journey, which is better worth remembering than anything else of all that occurred in it.

I.

But little original research was done in 1846 and 1847.

The experiments in the laboratory in 1846 were chiefly with regard to the action of magnetism on a polarised ray of light. At the end of the year a number of experiments were made on the purity of ice.

He published, in the 'Philosophical Magazine,' his

thoughts on ray vibrations, and on the magnetic affection of light; and in the 'Bibliothèque Universelle,'
letters to MM. de la Rive and Dumas, on the influence
of magnetism on light.

In 1847, but little laboratory work was done until
the middle of October. Some experiments were made
on the purity of ice, and some on the action of magnetism
upon light, and on the decomposition by electricity of
the iodide of nitrogen; but no important results were
obtained. At the end of the year the results of his experiments on the diamagnetic conditions of flame and
gases were sent to the 'Philosophical Magazine.' The
discovery had been made by Bancalari, Professor at
Genoa. Faraday says, 'I scarcely know how I could
have failed to observe the effect years ago.' The experiments were extended, and differential actions examined,
and he considered that he announced the discovery of
the magnetic property of oxygen when he wrote : 'The
attraction of iron filings to a magnetic pole is not more
striking than the appearance presented by the oxygen
in coal-gas.' In consequence of the great difference in
respect of the magnetic relation between oxygen and
nitrogen, he attempted to separate air by magnetic
force alone into its chief constituents.

He had a paper in the 'Philosophical Magazine' on
the diamagnetic conditions of flame; and in the Report
of the British Association he published a discourse on
the magnetic condition of matter.

For the Institution, in the spring of 1846, he gave a
course of eight lectures on electricity and magnetism,
their differences and their unity. In his fifth lecture he
said of Oersted's discovery of the magnetic influence
of electricity in motion, 'It burst open the gates of a

domain in science, dark until then, and filled it with a
flood of light.' He might have used the same words
regarding his own discovery of magneto-electricity.

Early in the year he gave a Friday discourse on the
relation of magnetism and light. In his last note for
this lecture he writes, ' Perhaps hereafter obtain magnetism from light.' He gave another on the magnetic
condition of matter; and, later in the season, another
on Wheatstone's electro-magnetic chronoscope, at the
end of which he said he was ' induced to utter a speculation long on his mind, and constantly gaining strength
—viz., that perhaps those vibrations by which radiant
agencies, such as light, heat, actinic influence, &c.,
convey this force through space, are not vibrations of
an ether, but of the lines of force which, in this view,
equally connect the most distant masses together, and
make the smallest atoms or particles by their properties
influential on each other and perceptible to us ; ' and in
his notes he says, ' so incline to dismiss the ether.' A
little later he sends these views to the ' Philosophical
Magazine ' as thoughts on ray vibrations : ' But, from
first to last, understand that I merely throw out, as
matter for speculation, the vague impressions of my
mind ; for I give nothing as the result of sufficient
consideration, or as the settled conviction, or even
probable conclusion, at which I had arrived.' His last
Friday discourse was on the cohesive force of water.

To the Secretary of the Institution, who consulted
him regarding evening lectures, he said : ' I see no
objection to evening lectures if you can find a fit man
to give them. As to popular lectures (which at the
same time are to be *respectable* and *sound*), none are
more difficult to find. Lectures which *really teach* will

never be popular; lectures which are popular will
never *really teach.* They know little of the matter
who think science is more easily to be taught or learned
than A B C; and yet who ever learned his A B C
without pain and trouble? Still lectures can (generally) inform the mind, and show forth to the attentive
man what he really has to learn, and in their way are
very useful, especially to the public. I think they
might be useful to us now, even if they only gave an
answer to those who, judging by their own earnest
desire to learn, think much of them. As to agricultural
chemistry, it is no doubt an excellent and a popular
subject, but I rather suspect that those who know least
of it think that most is known about it.'

In 1847, for the Institution, after Easter, he gave a
course of eight lectures on physico-chemical philosophy.

In his first lecture, speaking of oxygen, he says:
'It is, perhaps, most directly among secondary causes,
that substance by which *we live,* and *move,* and *have
our being.'* And in his second lecture he says: 'It may
move attention to point out that we ourselves are cases
of *instant* (constant) action, either whilst in life or after
death. What is our *constant breathing* for, but to
supply a substance which, performing at the instant its
chemical duty, has then to be removed and replaced by
more? What is *our food* for, but to supply by a series
of instant chemical changes the other acting substance
in a fit state, and at the fitting place for this action to
go on? What is *our* death and dispersion to the
elements but the results of the action of that chemical
force amongst the materials of our body and circum-
jacent matter which succeeds to the previous living
chemistry so soon as the breath of our nostrils is

stopped? Guard against the supposition that this is all; but we have a right to show how much it is.'

He ended this course of lectures thus: ' In conclusion, I may remark that, whilst considering the state and condition of the powers with which matter is endowed, we cannot shut out from our thoughts the consequences as far as they are manifested to us, for we find them always for our good; neither ought we to do so, for that would be to make philosophy barren as to its true fruits. And when we think of the way in which heat and cohesive force are related to each other, and learn also that the sun is continually giving to the surface of this globe of ours warmth equal to the combustion of sixty sacks of coals in twelve hours on each acre of surface in this climate in the average of the year; and find that of the bodies thus heated, those which ought to remain solid are so circumstanced that they do remain solid; and those which, like *oxygen* and *nitrogen*, in our atmosphere ought to remain gaseous *do* remain gaseous; when we find, moreover, that wonderful substance, water, assuming under the same influence the solid, the fluid, and the gaseous states at natural temperatures, and so circulating through the heavens and the earth ever in its best form; and perceive that all this is done by virtue of powers in the molecules which are indestructible, and by laws of action the most simple and unchangeable, we may well, if I may say it without irreverence, join *awe and trembling* with *joy and gladness.*

' Our philosophy, feeble as it is, gives us to see in every particle of matter, a *centre* of force reaching to an infinite distance, binding worlds and suns together, and unchangeable in its permanency. Around this

same particle we see grouped the powers of all the 1846–47.
various phenomena of nature : the heat, the cold, the Æt.54–56.
wind, the storm, the awful conflagration, the vivid
lightning flash, the stability of the rock and the mountain, the grand mobility of the ocean, with its mighty
tidal wave sweeping round the globe in its diurnal
journey, the dancing of the stream and the torrent;
the glorious cloud, the soft dew, the rain dropping
fatness, the harmonious working of all these forces in
nature, until at last the molecule rises up in accordance
with the mighty purpose ordained for it, and plays its
part in the gift of *life itself.* And therefore our
philosophy, whilst it shows us these things, should
lead us to think of Him who hath wrought them; for
it is said by an authority far above even that which
these works present, that " the invisible things of Him
from the creation of the world are clearly seen, being
understood by the things that are made, even His
eternal power and Godhead." '

He gave Friday evenings on the combustion of gunpowder; on Mr. Barry's mode of ventilating the new
House of Lords; and on the steam-jet chiefly as a
means of procuring ventilation.

For the Trinity House he did little work during
these years.

He reported, in 1846, on the drinking-water of the
Smalls Lighthouse, and on a ventilation apparatus for
rape-oil lamps.

He reported, in 1847, on the ventilation of the South
Foreland lights, and on a proposal to light buoys by
platinum wire ignited by electricity.

II.

In 1846, for his two great discoveries, the Rumford and the Royal Medals were both awarded to him. This double honour will probably long be unique in the annals of the Royal Society. In former years he had already received the Copley and Royal Medals for his experimental discoveries. As his medals increased it became remarkable that he—who kept his diploma-book, his portraits and letters of scientific men, and everything he had in the most perfect order—seemed to take least care of his most valuable rewards. They were locked up in a box, and might have passed for old iron. Probably he thought, as others did afterwards, that their value, if seen, might lead to their loss.

He was made Honorary Member of the Vaud Society of Sciences.

In 1847 he was made member of the Academy of Sciences, Bologna ; Foreign Associate of the Royal Academy of Sciences, Belgium ; Fellow of the Royal Bavarian Academy of Sciences, Munich ; Correspondent of the Academy of Natural Sciences, Philadelphia.

III.

His character is strongly shown in the letters he wrote regarding the losses that happened in these years in his family ; also in some letters on business for the Admiralty, and in a letter to M. Matteucci, and another to Professor Schönbein.

In August, 1846, his brother, when driving, was upset, and so injured in the head that he did not recover his consciousness again. He was taken to University College Hospital.

Faraday writes to Mrs. Faraday at Tunbridge Wells :

FARADAY TO MRS. FARADAY.

'Royal Institution : Thursday, August 13, 1846.

' Dear Heart,—My brother, exhausted by the results of the terrible accident which happened to him on Tuesday night, died this morning about seven o'clock. He had spent the evening of Tuesday at Mr. B.'s. After supper, he brought home my sister in the cart, and then proceeded to take the horse and cart to the stables. It is supposed the horse must have been frightened, or run against a post, but the cart was over-turned, my poor brother cast out, and so injured on the head as not to recover his consciousness again. He was taken to University Hospital, where I saw his corpse this morning, and though sadly bruised, it was just my dear brother.

.   .   .   .   .   .

' Come home, dear. Come and join in the sympathy and comfort needed by many. Let the establishment at Tunbridge Wells go on, and give my love to father and mother, and Margery, who will keep it up for a while for us.

' My sister and her children have not forgotten the hope in which they were joined together with my dear Robert, and I see its beautiful and consoling influence in the midst of all these troubles. I and you, though joined in the same trouble, have part in the same hope.

' Come home, dearest. Your affectionate husband,
'M. FARADAY.'

The words of feeling and of sense which he wrote to his nephew who had lost his wife in childbirth give a

1846–47.
Ær.54–56.

picture of Faraday which may be well called perfect. 'If the loss be sudden and grievous to us, how much more so must it be to you; and indeed we feel *deeply* for you. Let us hope and think that strength will be given you to bear it as a man; patiently, as one to whom grief and adversity does not come bringing only distress and unavailing sorrow, but deeper thoughts and instruction which afterwards produces good fruit. There are none of us who do not need such teaching, but it is hard to bear; and indeed, my dear J., it is *very hard* when it comes with such an overwhelming flood as that which has just reached you. I know that no words of mine are fitted to comfort you, but I seek only to sympathise, and you may believe how earnestly I do so when I at the same moment think what my state would have been had your loss been mine.

'Give a kiss to the children from me. Remember me to your mother, and think of me as your very affectionate uncle,　　　　　'M. FARADAY.'

He thus writes to the Right Honourable Lord Auckland, then First Lord of the Admiralty.

FARADAY TO LORD AUCKLAND.

'14 Duke Street, Edinburgh: July 29, 1847.

'My dear Lord,—If I had been in London I should have waited on your Lordship at the Admiralty instantly; as it is, I can only express my readiness to have done so. In reference to your Lordship's request, I will now take the liberty of explaining my position, which I did not very long since in a letter to the Secretary of the Admiralty, to which letter, however, and to a former

one, containing the results of serious consideration and 1846–47.
much time, I have never received any reply.　　　　　Ӕᴛ.54–56.

'For years past my health has been more and more
affected, and the place affected is my head.　My
medical advisers say it is from mental occupation.
The result is loss of memory, confusion and giddiness;
the sole remedy, cessation from such occupation and
head-rest.　I have in consequence given up, for the
last ten years or more, all professional occupation, and
voluntarily resigned a large income that I might pursue
in some degree my own objects of research.　But in
doing this I have always, as a good subject, held
myself ready to assist the Government if still in my
power—*not for pay*, for, except in one instance (and
then only for the sake of the person joined with me), I
refused to take it.　I have had the honour and pleasure
of applications, and that very recently, from the Admiralty, the Ordnance, the Home Office, the Woods and
Forests, and other departments, all of which I have
replied to, and will reply to as long as strength is left
me; and now it is to the condition under which I am
obliged to do this that I am anxious to call your Lordship's attention in the present case.　I shall be most
happy to give my advice and opinion in any case as
may be at the time within my knowledge or power,
but I may not undertake to enter into investigations or
experiments.　If I were in London I would wait upon
your Lordship, and say all I could upon the subject of
the disinfecting fluids, but I would not undertake the
experimental investigation; and in saying this I am
sure that I shall have your sympathy and approbation
when I state that it is now more than three weeks
since I left London to obtain the benefit of change of

air, and yet my giddiness is so little alleviated that I
don't feel in any degree confident that I shall ever be
able to return to my recent occupations and duties.

'I have the honour to be, my Lord, your Lordship's
very faithful servant,

'M. FARADAY.'

LORD AUCKLAND TO FARADAY.

'Admiralty : August 1, 1847.

'My dear Sir,—I have received your letter, and
return you many thanks for it.   I am very sorry for the
account which you give me of your health, and must
express my hope that the quiet which you are seeking
may completely restore it.   For the investigation which
I have immediately in view I will look for other assistance.   In the event of your returning to us in perfect
health I would again occasionally recur to your friendly
advice.

'Very truly, &c.,

'AUCKLAND.'

He writes to Capt. W. A. B. Hamilton, R.N., Secretary, Admiralty.

'Royal Institution : August 30, 1847.

'Sir,—I regret that I cannot at present accede to the
request[1] contained in your letter of the 26th instant.
The state of my health is one reason, but there are
others which I am too weary to write you at present,
but you will find them in a letter which I wrote to Mr.
Ward about three months ago, and to which I beg to
direct the attention of my Lords Commissioners and

---

[1] To report upon the working of Mr. Gamble's electro-magnetic telegraph upon the Great Western Railway.

yourself. I may be allowed to express a little surprise 1846–47.
that I have not yet received any reply, either to that Ær.54-56.
letter or to my former report (in conjunction with
Capt. Brandreth) on this particular telegraph.

' I have the honour to be, Sir, your most obedient,
humble servant,

' M. FARADAY.'

The following note, written sixteen years after his
' unfortunate letter ' to M. Hachette, shows that he was
entirely free from all jealousy of others as regarded
their work on the subjects that he might well have
looked upon as his own property.

FARADAY TO C. MATTEUCCI.

'Royal Institution : May 29, 1847.

' My dear Matteucci,—I find you are at work on
specific induction.

' My ideas are very obscure on the matter now, for
I cannot recollect the facts, and shall avoid doing so
until you have worked out your views, and very likely
for a good while after, for I am never better pleased
than to see others work on points I have dealt with,
and the points themselves either enlarged or multiplied
or corrected.

' Most truly yours,
' M. FARADAY.'

FARADAY TO PROFESSOR SCHÖNBEIN.

'Royal Institution: Oct. 23, 1847.

' My dear Schönbein,—With absolutely nothing to
say, I still feel a lingering desire to write to you, and
though I have waited days and weeks in hopes that
my thoughts would brighten, I will wait no longer, but

just make a return to your very characteristic letter by one which will be distinguished only by its contrast with it. You would perhaps see by the " Philosophical Magazine " that I had received yours, for the whole was printed there except three or four lines at the end. The novelty and beauty of your new test for ozone is very remarkable, and not less its application to the detection of the ozone from such different sources as phosphorus, the electrical brush, and electrolysis. I shame to say that I have not yet repeated the experiments, but my head has been so giddy that my doctors have absolutely forbidden me the privilege and pleasure of working or thinking for a while, and so I am constrained to go out of town, be a hermit, and take absolute rest. In thinking of my own case it makes me rejoice to know of your health and strength, and to look on whilst you labour with a constancy so unremitting and so successful. Long may it be so, to the joy and happiness of yourself, wife, and family.

.   .   .   .   .   .

'Ever, my dear Schönbein, yours most truly,

'M. FARADAY.

'I do not talk about gun-cotton, because I think you will let me know when anything philosophical or important turns up respecting it which it would give you pleasure to tell me. But you may suppose that I do not hope the less in respect of it.'

The years 1848–1849 give no strong lines in the life of Faraday.

I.

The early part of 1848 but little work was done in the laboratory. In May, some experiments were made with a view to the retardation effect of solar rays in combustion. In August, Professor Plücker was here, and described to Faraday certain of his results as to crystalline diamagnetic relation. Three days in August, Plücker was in the laboratory; and all September and October, Faraday worked on the crystalline polarity of bismuth, and on its relation to the magnetic force. From this resulted the twenty-second series of 'Experimental Researches.' The first part was sent to the Royal Society, October 4; and the second, October 31.

Dr. Tyndall says of this investigation :—

'The action of crystals had been in part theoretically predicted by Poisson, and actually discovered by Plücker, whose beautiful results, at the period which we have now reached, profoundly interested all scientific men. Faraday had been frequently puzzled by the deportment of bismuth, a highly crystalline metal. Sometimes elongated masses of the substance refused to set equatorially, sometimes they set persistently oblique, and sometimes even, like a magnetic body, from pole to pole. " The effect," he says, " occurs at a single pole; and it is then striking to observe a long piece of a substance so diamagnetic as bismuth repelled, and yet at the same moment set round with force, axially, or end on, as a piece of magnetic substance would do." The phenomena, he concludes, are altogether different from those of magnetism or diamagnetism; they would appear, in fact, to present to us " a new force, or a new form of force, in the molecules of matter," which for convenience' sake he designates by a new word, as " the *magne-crystallic* force."

'After the description of the general character of this new force, Faraday states with the emphasis here reproduced its mode of action : " The *law* of action appears to be that *the line or axis of* MAGNE-CRYSTALLIC *force* (being the resultant of the action of all the molecules) *tends to place itself parallel, or as a tangent to the magnetic curve, or line of magnetic force passing through the place where the crystal is situated.*" The magnecrystallic force, moreover, appears to him "to be clearly distinguished from the magnetic or diamagnetic forces, in that it causes neither approach nor recession, consisting not in attraction or repulsion, but in giving a certain determinate position to the mass under its influence." And then he goes on "very carefully to examine and prove the conclusion that there was no connection of the force with attractive or repulsive influences."

'In conclusion. " This force," he says, " appears to me to be very strange and striking in its character. It is not polar, for there is no attraction or repulsion." And then, as if startled by his own utterance, he adds :—" What is the nature of the mechanical force which turns the crystal round, and makes it affect a magnet ? " . . . " I do not remember," he continues, " heretofore such a case of force as the present one, where a body is brought into position only, without attraction or repulsion."

'At the end of his papers, when he takes a last look along the line of research, and then turns his eyes to the future, utterances quite as much emotional as scientific escape from Faraday. " I cannot," he says, at the end of his first paper on magne-crystallic action, " conclude this series of researches without remarking

how rapidly the knowledge of molecular forces grows
upon us, and how strikingly every investigation tends
to develope more and more their importance, and their
extreme attraction as an object of study.   A few years
ago, magnetism was to us an occult power, affecting
only a few bodies; now it is found to influence all
bodies, and to possess the most intimate relations with
electricity, heat, chemical action, light, crystallisation, and through it, with the forces concerned in cohesion ; and we may, in the present state of things,
well feel urged to continue in our labours, encouraged
by the hope of bringing it into a bond of union with
gravity itself." '

Professor Tyndall continues :—

' Plücker's and Faraday's investigations filled all
minds at the time, and towards the end of 1849,
Professor Knoblauch and myself commenced a joint
investigation of the entire question.   Long discipline
was necessary to give us due mastery over it.   Employing a method proposed by Dove, we examined the
optical properties of our crystals ourselves ; and these
optical observations went hand in hand with our
magnetic experiments.   The number of these experiments was very great, but for a considerable time no
fact of importance was added to those already published.
At length, however, it was our fortune to meet with
various crystals whose deportment could not be brought
under the laws of magne-crystallic action enunciated
by Plücker.   We also discovered instances which led
us to suppose that the magne-crystallic force was by
no means independent, as alleged, of the magnetism or

diamagnetism of the mass of the crystal. Indeed, the more we worked at the subject, the more clearly did it appear to us that the deportment of crystals in the magnetic field was due, not to a force previously unknown, but to the modification of the known forces of magnetism and diamagnetism by crystalline aggregation.

'These and numerous other results bearing upon the question were published at the time in the "Philosophical Magazine" and in Poggendorff's "Annalen;" and the investigation of diamagnetism and magne-crystallic action was subsequently continued by me in the laboratory of Professor Magnus of Berlin.

'It required long subsequent effort to subdue the complications of magne-crystallic action, and to bring under the dominion of elementary principles the vast mass of facts which the experiments of Faraday and Plücker had brought to light.

'The most perplexing of those facts were shown to result from the action of mechanical couples, which the proved polarity both of magnetism and diamagnetism brought into play. Indeed, the thoroughness with which the experiments of Faraday were thus explained, is the most striking possible demonstration of the marvellous precision with which they were executed.'

In the early part of 1849 the laboratory work consisted of experiments on the possible relation of gravity to electricity. In September, October and November the polarity of bismuth, copper, phosphorus in the magnetic field was again examined. Again he arrived only at negative results, but these he sent to the Royal Society as the twenty-third series of 'Experimental Researches' on the polar or other condition of diamagnetic bodies.

In the Edinburgh new 'Philosophical Journal' he had a paper in 1848 on the use of gutta-percha in electrical insulation.

For the Institution in 1848 he gave a course of seven lectures on the allied phenomena of the chemical and electrical forces.

In his notes on the voltaic battery, he wrote : ' Here consider how Davy wielded this power, and triumphed over the alkalies in this room.' And of the combustion of the zinc, he writes : 'And this gives one a strange sensation as to what may be going on in a gas flame or a fire ; ' and hopes that some day we may transfer their light and heat and all their other powers to a distance, and use them at pleasure, laying on, not gas, but the powers of the gas or oil, and so having a lamp more wonderful than Aladdin's.

He ended his course thus : ' I would rather leave you admiring with me the wonderful way in which we thus see the same force made manifest under such extraordinary differences, than whilst I see but the clouded truth pretend to explain it clearly. We are all in the condition of those who, being thirsty, have not that which is needed, and therefore should conjointly desire and struggle for the knowledge that *is* to be revealed.'

In 1849, at the Institution, after Easter, he gave eight lectures on static electricity. Early in the first lecture he says in his notes : ' What an idea of the ever present and ever ready state of this power is given to us, when we thus consider that not only *every substance*, but almost every mode of dealing with a substance, manifests its presence. It is not accidental at these times, but active, and essentially so; and we may in our endeavours to comprehend it usefully compare and contrast it with gravity, which never changes. There we see that power which in *undisturbed* and solemn grandeur holds equally the world and the dust of which worlds are formed together, and carries them on in their course through illimitable space for illimitable ages; and in *this other* power, even in this our first glimpse, we see probably the *contrasted* force which is destined to give all that vivacity and mutual activity to particles that shall fit them, as far as matter alone is concerned, for their wonderful office in the phenomena of Nature, and enable them to bring forth the ever varying and astonishing changes which earth, air, fire, and water present to us, from the motion of the dust in the whirlwind up to the highest conditions of life.'

By accident during this course he received a severe shock from the Leyden battery. He laughed hysterically, and the audience, thinking that he laughed jokingly, joined in the joke. With some difficulty he continued the lecture.

He gave three Friday discourses. On the diamagnetic condition of flame and gases; on two recent inventions of artificial stone; and on dimorphism or allotropic conditions, the conversion of diamond into coke by the electric flame.

In 1849 he gave three Friday discourses. On magnecrystallic phenomena; on Plücker's repulsion of the optic axis of crystals by the magnetic poles; and on De la Rue's envelope machinery.

On Monday morning, February 26, Prince Albert came to the Institution to a private lecture on magnetic

and diamagnetic bodies.   Faraday's notes begin thus :
'The exertions in physical science of late years have
been directed to ascertain, not merely the different
natural powers, but the manner in which they are
linked together, the universality of each in its action,
and their probable *unity in one.*'   He ends his notes
thus : 'Such are the recent additions to our knowledge
respecting the manner in which the magnetic force
exercises its governing power over matter ; and when
we thus see it extended to all matter, animal, vegetable,
or mineral, living or lifeless, and when we see it thus
making strange or striking distinctions between the same
kind of matter, as it may be in the amorphous or crystalline state, and when we remember that the earth itself is
a magnet, pervaded in every part by this mighty power,
universal and strong as gravity itself, we cannot doubt
that it is exerting an appointed and essential influence
over every particle of matter and in every place where
it is present.   What its great purpose is, seems to be
looming in the distance before us, the clouds which
obscure our mental sight are daily thinning, and I
cannot doubt that a glorious discovery in natural
knowledge, and of the wisdom and power of God in
the creation, is awaiting our age, and that we may not
only hope to see it, but even be honoured to help in
obtaining the victory over present ignorance and future
knowledge.'

He held the same hope unchanged when in his
seventieth year he wrote : 'Nothing would make me
happier in the things of this life, than to make some
scientific discovery or development.'

He gave the Juvenile Lectures on the chemical
history of a candle.

He reported to the Trinity House in 1849 on the ventilation of Flambro' Head, Dungeness, Needles, and Portland Lighthouses.

II.

His reputation was marked in the year 1848 by two titles.

He was made Foreign Honorary Member (one of eight) of the Imperial Academy of Sciences, Vienna, and Doctor of Liberal Arts and Philosophy in the University of Prague ; and a letter from M. Dumas is of some interest as evidence of his regard for Faraday.

In 1849 he was made Honorary Member, First Class, Institut Royale des Pays-Bas, and Foreign Correspondent of the Institute, Madrid.

A letter from Professor Oersted, the discoverer of electro-magnetism, speaks of the value of Faraday's researches.

M. DUMAS TO FARADAY.

'A la Sorbonne, le 24 juillet, 1848.

'Mon cher ami,—Les événements si tristes, mais, hélas ! si bien prévus, que Paris a dû subir n'ont affecté ni moi ni les miens d'une manière directe.   Mdme. Dumas et moi, nous avons été bien touchés de votre marque de bonté et de souvenir ; c'est une consolation que cette affection des âmes élevées, comme la vôtre, au milieu d'un désordre moral dont rien n'en peut vous donner une idée, et qu'aucune imagination n'aurait certainement pu soupçonner.

'Paris était devenu, depuis six mois, le rendez-vous de tous les scélérats de la France et de l'Europe.   Les uns comme chefs, les autres comme instruments, tous

ensemble ils s'étaient proposé le pillage, le meurtre, 1848.
l'incendie, et tous les désordres comme les moyens de Æt.56–57.
régénérer notre nation, en détruisant sa bourgeoisie et
en livrant tous les pouvoirs, toutes les fortunes, et toutes
les familles au despotisme et à la brutalité des classes
ouvrières.

'Tout est voilé ici ; les arts, les lettres, les sciences,
tout se ressent du deuil universel. Les fortunes, les
prolétaires, les existences, tout a été mis en question
par les événements qui se sont succédés.

'M. Arago me charge de vous dire combien votre
bon souvenir lui a été doux au cœur. Sa conduite
dans ces dernières journées de péril a été si ferme et si
courageuse, qu'il s'est jeté sur les barricades au milieu
des balles avec tant de résolution, que les personnes
qui ont pu le voir dans ces circonstances ont pu croire
qu'il cherchait une mort glorieuse, désespérant du salut
du pays. Il faut bien convenir que nous en étions tous
là, et que cette triste pensée ne pouvait guère s'éloigner
de nos cœurs quand nous songions à l'immensité des
ressources des insurgés et à la faiblesse des moyens de
résistance que nous possédions.

'Que Dieu vous préserve, mon cher ami, votre pays
et vous, de ces lamentables folies ! Jamais nous ne
cicatriserons les plaies ouvertes et envenimées depuis
quelques mois par la presse, les clubs, les sociétés
secrètes, et surtout les ateliers nationaux. Partout la
haine de toutes les supériorités, la soif de toutes les
richesses, la méprise de tout ce que l'homme doit
respecter ; voilà ce qui a fait la base des écrits, du
discours et des associations.

'J'interromps ma lettre pour lire le billet qui m'annonce la mort d'un ami, blessé il y a un mois. Tous

les jours il en est ainsi ; les convois se succèdent, et nous sommes loin d'avoir fini ce compte funèbre.

'Adieu, mon cher ami. Pardonnez-moi si je vous réponds si tard, mais votre lettre ne m'est pas parvenue comme elle aurait dû. J'ai changé de logement. Me voici à la Sorbonne ; et puis je suis bien découragé, bien triste.

'Mes respects à Madame Faraday, je vous prie.

'Mille amitiés.

'Dumas.'

PROFESSOR OERSTED TO FARADAY.

'Copenhagen : December 27, 1849.

'My dear Sir,—Permit me to recall myself in your remembrance, in introducing to the honour of your acquaintance the bearer of this letter, Mr. Colding.

. . . . .

'You have, in earlier years, had the kindness to send me your series of "Researches," of which I possess the first seventeen. Though I have the whole series in the "Transactions," I should be glad to be in possession of the continuation of these immortal papers. I consider already those which I have as one of the most distinguished ornaments of my library.

'I am, dear Sir, with sincere admiration, most faithfully yours,

'H. C. Oersted.'

III.

His love of quiet and of home is seen in a letter to Mrs. Faraday from Birmingham ; and the conclusion of a letter to Mr. Brodie shows how he estimated his own position.

In July he wrote to the Rev. John Barlow, from Clevedon, Somersetshire.

FARADAY TO REV. J. BARLOW.

' Your kind letter came safe to hand, and was very welcome. It finds my head so confused that whether I shall spell or write my words rightly I really cannot tell. I knew not until I made up my mind to rest, how much I wanted it; and should think that I was more feeble now than when I left London, did I not know from old experience that I have first to sink to a natural state before I can naturally and healthily rise from it. I expect, however, that I shall obtain what I want, though not perhaps as soon as I desire. If, therefore, I should feel that a week longer here would complete the good I want, I think I may stop.'

FARADAY TO W. R. GROVE, ESQ.

(who had asked for some heavy glass for Matteucci).

' Royal Institution : November 9, 1848.

' My dear Grove,—Matteucci's embarrassment results, I suppose, from his having already had some heavy glass, but though I have none to waste, I have as yet plenty for all who want it for use. I send you herewith a piece for him, with my kindest remembrances. I wrote to him by the Marquis Ridolfi only a week or two ago ; else I would punish him with a letter. I do not know whether you have or would care for a piece of the heavy glass. I have sent another plate to be cut into pieces. It is not so good as the glass already dispersed, but I have no doubt will do, except that it may be deficient in the annealing. However, that does not prevent, though it modifies,

R 2

the optic phenomena. When it comes back you shall
have a piece, if you like ; only I am afraid I may FORGET,
for that is my great and continual trouble.
'Ever yours,
'M. FARADAY.'

FARADAY TO MRS. FARADAY.

'Birmingham, Dr. Percy's:
'Thursday evening, September 13, 1849.

'My dearest Wife,—I have just left Dr. Percy's
hospitable table to write to you, my beloved, telling
you how I have been getting on. I am very well,
excepting a little faceache; and very kindly treated
here. They all long most earnestly for your presence,
for both Mrs. and Dr. Percy are anxious you should
come ; and this I know, that the things we have seen
would delight you, but then I doubt your powers of
running about as we do; and though I know that if
time were given you could enjoy them, yet to press
the matter into a day or two would be a failure.
Besides this, after all, there is no pleasure like the
tranquil pleasures of home, and here—even here—the
moment I leave the table, I wish I were with you IN
QUIET. Oh ! what happiness is ours ! My runs into
the world in this way only serve to make me esteem
that happiness the more. I mean to be at home on
Saturday night, but it may be late first, so do not be
surprised at that ; for if I can, I should like to go on an
excursion to the Dudley caverns, and that would take
the day.
'Mr. Daniel called on me to-day with a pressing
invitation for you and me to his house, for which I

thanked him sincerely, as he deserved to be thanked, but I could give no hopes of that.

'Write to me, dearest. I shall get your letter on Saturday morning, or perhaps before.

'Love to father, Margery, and Jenny, and a thousand loves to yourself, dearest,

'From your affectionate husband,

'M. FARADAY.'

FARADAY TO DR. PERCY.

'Royal Institution: October 17, 1849.

'My dear Percy,—I cannot be on the Committee; I avoid everything of that kind, that I may keep my stupid mind a little clear. As to being on a Committee and not working, that is worse still.

'I wish we could get to Birmingham, and use your kindness, but that may not be. My working time is from October to December, and I am fully in it: I am sorry to say, as yet, with negative results. Still I must work.

'Ever yours and Mrs. Percy's,

'M. FARADAY.'

FARADAY TO C. MATTEUCCI.

'Royal Institution: November 5, 1849.

'My dear Matteucci,— . . . . I have lately been working for full six weeks trying to procure results, and have indeed procured them, but they are all negative. But the worst of it is, I find on looking back to my notes, that I ascertained all the same results experimentally eight or nine months ago, and had entirely forgotten them. This in some degree annoys me. I do not mean the labour, but the forgetfulness, for, in fact, the labour without memory is of no use.

'Still I have a thousand causes of thankfulness, and am not repining, only explaining. If I could have my own way, I would never write you a letter without some scientific point in it. As it is, the chances are they will be as barren as this one.

'Yours most truly,

'M. FARADAY.'

FARADAY TO B. C. BRODIE, ESQ.

'Royal Institution: December 17, 1849.

'My dear Brodie,—I owe you many and sincere thanks for your kind note. As to your letter to the Secretary, which was of course read to the Managers, it contained so absolute a negative on your part to their request to give another course of lectures at the Institution, that everybody felt there was no more to say upon the matter. The Secretary might, and very probably by this time has, acknowledged the receipt of it.

'And now, my dear Sir, though it was this affair that chiefly made you and me known to each other, and though it has ended otherwise than I hoped, still I shall not, as regards ourselves, let matters return to their former state.

'I hope much from you, and shall, as long as I remain in life, look with expectation, and, I trust, rejoicing, to your course. If any word from me is of the least value as a word of encouragement and exhortation, I say *proceed, advance.*

'Here things have reverted very much to their former state, I rather think perhaps fitly. The time

was probably too soon for any change. But when
such an one as myself gets out of the way, then new
conditions, new men, new views, and new opportunities
may allow of the development of other lines of active
operation than those heretofore in service ; and then
perhaps will be the time for change.

'Ever, my dear Sir, very truly yours,

'M. FARADAY.'

I.

In 1850 more original research was done than in any
year since his illness. He had no less than five papers
read at the Royal Society. The first of these was on the
polar condition of diamagnetic bodies. The second
was on the possible relation of gravity to electricity.

On March 19, 1849, he writes in the laboratory
book thus : ' Gravity. Surely this force must be capable of an experimental relation to electricity, magnetism,
and the other forces, so as to bind it up with them in
reciprocal action and equivalent effect. Consider for
a moment how to set about touching this matter by facts
and trial.

' What in gravity answers to the dual or antithetical
nature of the forms of force in electricity and magnetism ?
Perhaps the *to* and *fro*, that is, the ceding to the force
or approach of gravitating bodies, and the effectual
reversion of the force or separation of the bodies, quiescence being the neutral condition. Try the question
experimentally on these grounds, then the following suppositions or suggestions arise.

' Bodies approaching by gravitation, and bodies separated per force whilst gravitating towards each other,
may show in themselves or in surrounding matter or

helices *opposite currents* of electricity round the line of motion as an axis. But if not moving *to* or *from* each other should produce no effect.'

And then he continues for twenty paragraphs noting subjects for examination and thought, and ends saying : ' ALL THIS IS A DREAM. Still examine it by a few experiments. Nothing is too wonderful to be true, if it be consistent with the laws of nature ; and in such things as these, experiment is the best test of such consistency.'

Again, March 28, he writes : ' If there should happen to be any result of the kind imagined, then a body moving *up* would produce one current, and moving *down* the reverse current. Now, these may be converted by a commutator into one consistent current, and that may be sent through a galvanometer for the time of a *half vibration* of the needle, and then by a *second* commutator be sent for the second half vibration in the contrary direction, then back in the first direction, and so on continually. This would seem to be a good way of accumulating the induced force or current, IF THERE BE ANY.'

The experiments which he made on this subject are recorded in the twenty-fourth series of ' Experimental Researches,' received by the Royal Society, August 1, 1850, on the possible relation of gravity to electricity. He finishes the paper thus : ' Here end my trials for the present. The results are negative ; they do not shake my strong feeling of the existence of a relation between gravity and electricity, though they give no proof that such a relation exists.' Ten years afterwards he says the same thing, almost in the same words, in the very last paper that he wrote, so constant was he even in science when he had made up his mind.

The third, fourth, and fifth papers, being the twenty-
fifth, sixth, and seventh series of ' Researches,' were on
the magnetic state of gases. Throughout the whole
year, the diamagnetic or magnetic condition of gases was
the subject of the laboratory work. June 24, he writes,
after having tried soap bubbles with air and nitrogen :
' *Oxygen.* Here the effect was beautiful, the bubble being
pulled inwards with very considerable force in air and
looking exactly as if the oxygen were highly magnetic.
This was of course expected, and accords with all the
phenomena of the old time.' And then he says : ' If one
could tell how a bubble of hot gas would move in an
atmosphere of the same gas cold, but rarefied to the same
degree as the hot, it would be important, only I do not
see how that is to be done.'

On July 13 he devises a way of getting over his
difficulty thus : ' May compare a rare and a dense gas
in thin glass cylinders placed on opposite sides of axial
line in magnetic field ; ' and on July 16 he immediately
compares dense and exhausted oxygen ; and July 27
he writes : ' The results were beautiful, showing that as
the oxygen was rarefied up to a vacuum it became less
and less powerful in its tendency inwards. Hence can
hardly resist the conclusion that it is a magnetic body,
and powerfully magnetic, and that as it is rarefied it loses
some of this power. The loss seemed to me very like
as if in proportion to the removal of oxygen.'

Then he estimates the amount of paramagnetic force
in oxygen.

These and other results he sent to the Royal Society
as the twenty-fifth series of ' Researches.'

He ends his paper thus :—

1850.

Æt.58–59.

'It is hardly necessary for me to say here that this oxygen cannot exist in the atmosphere, exerting such a remarkable and high amount of magnetic force, without having a most important influence on the disposition of the magnetism of the earth as a planet, especially if it be remembered that its magnetic condition is greatly altered by variations in its density and by variations in its temperature. I think I see here the real cause of many of the variations of that force which have been and are now so carefully watched on different parts of the surface of the globe. The daily variation and the annual both seem likely to come under it; also very many of the irregular continual variations which the photographic process of record renders so beautifully manifest. If such expectations be confirmed, and the influence of the atmosphere be found able to produce results like these, then we shall probably find a new relation between the aurora borealis and the magnetism of the earth, namely, a relation established more or less through the air itself in connection with the space above it; and even magnetic relations and variations which are not as yet suspected may be suggested and rendered manifest and measurable in the further development of what I will venture to call atmospheric magnetism. I may be over sanguine in these expectations, but as yet I am sustained in them by the apparent reality, simplicity, and sufficiency of the cause assumed as it at present appears to my mind. As soon as I have sufficiently submitted these views to a close consideration and the test of accordance with observation, and where applicable with experiments also, I will do myself the honour to bring them before the Royal Society.'

On October 9 the twenty-sixth series, and on

November 19 the twenty-seventh series of 'Researches' were sent to the Royal Society.

In a letter he drew a picture of himself whilst writing these papers for the Royal Society.

His letter is dated August 24, from Upper Norwood, to a friend who asked him to stay in the country in the summer :—' I have kept your picture to look at for a day or two before I acknowledge your kindness in sending it. It gives the idea of a tempting place ; but what can you say to such persons as we are who eschew all the ordinary temptations of society ? There is one thing, however, society has which we do not eschew ; perhaps it is not very ordinary, though I have found a great deal of it, and that is kindness, and we both join most heartily in thanking you for it, even when we do not accept that which it offers. I must tell you how we are situated. We have taken a little house here on the hill-top, where I have a small room to myself, and have, ever since we came here, been deeply immersed in magnetic cogitations. I write and write and write, until three papers for the Royal Society are nearly completed, and I hope that two of them will be good if they justify my hopes, for I have to criticise them again and again before I let them loose. You shall hear of them at some of the Friday evenings ; at present I must not say more. After writing, I walk out in the evening, hand-in-hand with my dear wife, to enjoy the sunset ; for to me who love scenery, of all that I have seen or can see there is none surpasses that of Heaven. A glorious sunset brings with it a thousand thoughts that delight me.'

Earlier the same friend asked him, for the first time, to dinner. He writes from Brighton :—' Your note is a very kind one, and very gratefully received ; I wish on some accounts that nature had given me habits more fitted to thank you properly for it by acceptance than those which really belong to me. In the present case, however, you will perceive that our being here supplies an answer (something like a lawyer's objection) without referring to the greater point of principle. I should have been very sorry in return for your kindness to say *no* to you on the other ground, and yet I fear I should have been constrained to do so.'

At the end of the year he had an invitation from the Hon. Colonel Grey. ' If you could make it convenient to come down to Windsor any afternoon in the course of next week, it would give His Royal Highness great satisfaction to have the opportunity of having some conversation with you on this interesting subject (the magnetic properties of oxygen).'

A letter to M. de la Rive, written early in the following year, gives a summary of the results of his discoveries on atmospheric magnetism.

FARADAY TO M. DE LA RIVE.

' Royal Institution : February 4, 1851.

' My dear de la Rive,—My wife and I were exceedingly sorry to hear of your sad loss : it brought vividly to our remembrance the time when we were at your house, and you and others with you made us so welcome. What can we say to these changes but that they show by comparison the vanity of all things under the sun ? I am very glad that you have spirits to return to work again, for that is a healthy and proper employment of the mind under such circumstances.

' With respect to my views and experiments, I do 1850. not think that anything shorter than the papers (and Æт.58–59. they will run to a hundred pages in the " Transactions ") will give you possession of the subject ; because a great deal depends upon the comparison of observations in different parts of the world with the facts obtained by experiment and with the deductions drawn from them : but I will try to give you an idea of the root of the matter. You are aware that I use the phrase *line of magnetic force*, to represent the presence of magnetic force, and the direction (of polarity) in which it is exerted ; and by the idea which it conveys one obtains very well, and I believe without error, a notion of the distribution of the forces about a bar-magnet, or between near flat poles presenting a field of equal force ; or in any other case. Now, if circumstances be arranged so as to present a field of equal force, which is easily done, as I have shown by the electro-magnet, then if a sphere of iron or nickel be placed in the field, it immediately disturbs the direction of the lines of force, for they are concentrated within the sphere. They are, however, not merely concentrated but *contorted* ; for the sum of forces in any one section across the field is always equal to the sum of forces in any other section ; and therefore their condensation in the iron or nickel cannot occur without this contortion. Moreover, the contortion is easily shown by using a small needle (one-tenth of an inch long) to examine the field : for, as before the introduction of the sphere of iron or nickel, it would always take up a position parallel to itself ; afterwards it varies in position in different places near the sphere. This being understood, let us then suppose the sphere to be raised in temperature ; at a certain temperature it begins to lose its power of affecting the lines of magnetic force, and ends by retaining scarcely any; so that as regards the little needle mentioned above, it now stands everywhere parallel to itself within the field of force. This change occurs with iron at a very high temperature, and is passed through within the compass, apparently, of a small number of degrees: with nickel it occurs at much lower temperatures, being affected by the heat of boiling oil.

'Now take another step. Oxygen, as I showed above three years ago in the "Philosophical Magazine" for 1847, vol. xxxi. pp. 410, 15, 16, is magnetic in relation to nitrogen and other gases. E. Becquerel, without knowing of my results, has confirmed and extended them in his paper of last year, and given certain excellent measures. In my paper of 1847 I showed also that oxygen (like iron and nickel) lost its magnetic power and its ability of being attracted by the magnet when heated, p. 417; and I further showed that the temperatures at which this took place were within the range of common temperature; for the oxygen of the air, i.e. the air altogether, increased in magnetic power when cooled to 0° F. p. 406. Now I must refer you to the papers themselves for the (to me) strange results of the incompressibility (magnetically speaking) of oxygen and the inexpansibility of nitrogen and other gases; for the description of a differential balance by which I can compare gas with gas, or the same gas at different degrees of rarefaction; for the determination of the true zero, or point between magnetic and diamagnetic bodies; and for certain views of magnetic conduction and polarity. You will there find described certain very delicate experiments upon diamagnetic and very

weak magnetic bodies concerning their action on each
other in a magnetic field of equal force; the magnetic
bodies repel each other, and the diamagnetic bodies
repel each other: but a magnetic and a diamagnetic
body *attract* each other; and these results, combined
with the qualities of oxygen, as just described, convince
me that it is able to deflect the lines of magnetic force
passing through it just as iron or nickel is, but to an
infinitely smaller amount; and, that its power of
deflecting the lines varies with its temperature and
degree of rarefaction.

' Then comes in the consideration of the atmosphere
and the manner in which it rises and falls in temperature by the presence and absence of the sun.
The place of the great warm region nearly in his neighbourhood;—of the two colder regions which grow up
and diminish in the northern and southern hemispheres
as the sun travels between the tropics;—the effect of
the extra warmth of the northern hemisphere over the
southern;—the effect of accumulation from the action of
preceding months;—the effect of dip and mean declination at each particular station;—the effects that follow
from the noncoincidence of magnetic and astronomical
conditions of polarity, meridians, and so forth;—the
results of the distribution of land and water for any given
place;—for all these and many other things I must
refer you to the papers. I could not do them justice
in any account that a letter could contain, and should
run the risk of leading you into error regarding them.
But I may say that, deducing from the experiments and
the theory what are the deviations of the magnetic
needle at any given station, which may be expected as
the mean result of the heating and cooling of the

atmosphere for a given season and hour, I find such a general accordance with the results of observations, especially in the direction and generally in the amount for different seasons of the *declination* variation, as to give me the strongest hopes that I have assigned the true physical cause of those variations, and shown the *modus operandi* of their production.

'And now, my dear de la Rive, I must leave you and run to other matters. As soon as I can send you a copy of the papers I will do so, and can only say I hope that they will meet with your approbation. With the kindest remembrances to your son,

'Believe me to be, my dear friend, ever truly yours,

'M. FARADAY.'

He reported on the adulteration of white lead for the Trinity House.

He did as much work as usual for the Royal Institution.

He gave a Friday discourse on the electricity of the air; and another on certain conditions of freezing water. In his notes he says: 'Perfect expulsion of salts, acids, alkalies in freezing, and hence the pure condition of ice, and the same mass may by virtue of the solidifying power at places of contact be freezing at the inside and thawing at the outside, i.e. *freezing* and *thawing* at the *same temperature*—and even that the freezing process in the inside may be a thawing process to the outside by the evolution, conduction, and absorption of the heat concerned.'

After Easter, he gave a course of six lectures upon some points of domestic chemical philosophy—a fire, a candle, a lamp, a chimney, a kettle, ashes. In his

notes on a lamp, he says of gas: 'The system of its
manufacture and supply is wonderful, and to many
of us still alive is like the realisation of the most
extravagant imaginings of fairy land : yet how commonplace or slight is the impression it makes now.
Cannot we, who are urged to move forward by the
prizes set before us, consider what great things may be
coming over us, founded on the powers already manifest
in the natural creation, which yet, if presented to our
minds clearly and at once, would even in our instructed
days still appear as the wildest creations of fancy?
There is nothing more strange in the human mind than
the manner in which a little familiarity in thought
reduces the highest and the most important of truths to
the dull no-interest of every-day commonplace.' Nearly
his last note of the course is : ' May conclude with the
full conviction that the little that is known is a great and
wonderful indication of that which is to be known.'

Four letters to Professor Schönbein, chiefly regarding
a Friday evening discourse on ozone, show how he
worked for the Institution, and show also his kindness
and energy.

FARADAY TO PROFESSOR SCHÖNBEIN

(who was at this time wishing Faraday to bring the
subject of ozone before the members of the Royal
Institution).

'Royal Institution : May 11, 1850.

' My dear Schönbein,—But now my thoughts are on
ozone. I like your idea of an evening here, but it cannot be this season, for the arrangements are full. Yet
that in some degree suits me better, for though I should
like to give it, I am a slow man (through want of

VOL. II.   s

memory), and therefore require preparation. Now I shall look up your letters and re-read them, and also the papers; but let me pray you to send me a list of the experiments which you know to suit a large audience. Also, if you can, the references to the best French (or English) papers giving an account of its development and progress. Also your present view; also the best and quickest mode of making ozonised air, and such other information as I shall need. Probably other matters will arise before 1851, and I will get possession of it as we go along. If you come over here you shall give the subject yourself, i.e. if you can arrange and keep to time, &c.; if not, I must do my best. But every year I need more cramming, even for my own particular subjects. Now do not delay to send me the list of experiments because you suppose there is plenty of time, &c., but let me have them, that I may think over them during the vacation. I should like to do the matter to my own satisfaction; there are, however, very few things in which I satisfy myself now. I hoped to have had a paper to send you ere this, but Taylor is slow in printing.

. . . . . .

'Ever, my dear Friend, yours truly,

'M. FARADAY.'

He shortly mentions his own discoveries in the following letter to Professor Schönbein.

FARADAY TO PROFESSOR SCHÖNBEIN.

'Royal Institution : November 19, 1850.

'My dear Schönbein,—I wish I could talk with you instead of being obliged to use pen and paper. I have fifty matters to speak about, but either they are too trifling for writing, or too important; for what can one discuss or say in a letter? Where is the question and answer, and explanation that brings out clear notions in a few minutes? whilst letters only make them more obscure, because one cannot speak freely one's notions, and yet guard them merely as notions. But I am fast losing my time and yours too. I received your complimentary kindness, and like it the better because I know it to be as real as complimentary. Thanks to you, my dear friend, for all your feelings of good will towards me. The bleachings by light and air are very excellent. I see a report of part of your paper in the account of the Swiss Association, but not of the latter part. However, a friend has your paper in hand, and I hope to have the part about atmospheric electricity soon sent to me. I should be very *glad indeed* to have from any one, and above all from you, a satisfactory suggestion on that point. I know of none as yet.

'By the bye, I have been working with the oxygen of the air also. You remember that three years ago I distinguished it as a magnetic gas in my paper on the diamagnetism of flame and gases founded on Bancalari's experiment. Now I find in it the cause of all the annual and diurnal, and many of the irregular, variations in the terrestrial magnetism. The observations made at Hobarton, Toronto, Greenwich, St. Petersburg, Washington, St. Helena, the Cape of Good Hope, and Singapore, all appear to me to accord with and support my hypothesis. I will not pretend to give you an account of it here, for it would require some detail, and I really am weary of the subject. I have sent in three long papers to the Royal Society, and you shall have copies of them in due time, and reports probably much sooner in "Taylor's Magazine." I forwarded you packets immediately on the receipt of them.

'But now about ozone. I was in hopes you would let me have a list of points with reference to where I should find the account in either English or French journals, and also a list of about twenty experiments fit for an audience of 500 or 600 persons; telling me what sized bottles to make ozone by phosphorus in, the time, and necessary caution, &c. &c. My bad memory would make it a terrible and almost impossible task, to search from the beginning and read up; whereas you, who keep all you read or discover with the utmost facility, could easily jot me down the real points. If you refer to any such notes in your last letter when you ask me whether I have received a memoir on ozone, and some other things, then I have not received any such notes, and I cannot—indeed *I cannot*—remember about the memoir.

'I was expecting some such notes, and I still think you mean to send me them, and though I may perhaps not give ozone as an evening *before Easter*, still do not delay to let me have them, because I am slow, and losing much that I read of, have to imbibe a matter two or three times over, and if I do *ozone* I should like to do it well.

'My dear wife wishes to be remembered to you, and I wish most earnestly to be brought to Madame Schönbein's mind, though vaguely I cling to the remembrance of an hour or two out of Bâle at your house; and though I cannot recall the circumstances clearly to my mind, I still endeavour again and again to realise the idea.

'Ever, my dear Schönbein, most truly yours,

'M. FARADAY.'

To Professor Schönbein he again writes on atmospheric magnetism.

FARADAY TO PROFESSOR SCHÖNBEIN.

'Brighton: December 9, 1850.

' My dear Schönbein,—I have just read your letter
dated July 9, 1850, exactly *six months* after it was
written. I received the parcel containing it just as I
was leaving London, and I do not doubt it was in consequence of your moving upon the receipt of my last
to you a few weeks ago. Thanks, thanks, my dear
friend, for all your kindness. I have the ozonometer,
and the summary, and all the illustrative packages safe,
and though I have read only the letter as yet I see
there is a great store of matter and pleasure for me.

' As to your theory of atmospheric electricity, I am
very glad to see you put it forward; of course such a
proposition has to dwell in one's mind, that the idea
may be compared with other ideas, and the judgment
become gradually matured; for it is not like the idea
of a new compound which the balance and qualitative
experiments may rapidly establish. Still, as I study
and think over your account of ozone and isolated
oxygen, so I shall gradually be able to comprehend
and imbibe the idea. Even as it is, I think it is as
good as any, and much better than the far greater
number of hypotheses which have been sent forth as to
the physical cause of atmospheric electricity; and some
very good men have in turns had a trial at the matter:
in fact, the point is a very high and a very glorious
one; we ought to understand it, and I shall rejoice if
it is you that have hold of the end of the subject. You
will soon pull it clearly into sight.

'As I told you in my last, I must talk about atmospheric *magnetism* in my Friday evening before Easter, and I am glad that ozone will fall in the summer months, because I should like to produce some of the effects here. I think I told you in my last how that oxygen in the atmosphere, which I pointed out three years ago in my paper on flame and gases as so very magnetic compared with other gases, is now to me the source of all the periodical variations of terrestrial magnetism; and so I rejoice to think and talk at the same time of your results, which deal also with that same atmospheric oxygen. What a wonderful body it is.

'Ever, my dear Schönbein, yours faithfully,
'M. FARADAY.'

He again writes to Professor Schönbein on the subject of ozone :—

FARADAY TO PROFESSOR SCHÖNBEIN.

'Brighton : December 13, 1850.

'It will be very strange if I do not make your subject interesting. I have gone twice through the MS. and the illustrations. Both are beautiful. As soon as I reach home I shall begin to prepare for ozone, making and repeating your experiments. This morning I hung out at my window one of the ozonometer slips. That was about two hours ago. Now when I moisten it, a tint of blue comes out between Nos. 4 and 5 of the scale. Though I face the sea and have the wind on shore, still I am not aware that the spray can do this or anything that comes from the sea water; but before I send off this letter I shall go down and try the sea itself.

' Well, I have been to the seaside, and the sea water   1850.
does nothing of the kind, nor the spray ; but as I walk   Æt.58–59.
on the shore, holding a piece of the test-paper in my
hand for a quarter of an hour, at the end of that time
it, by moistening, shows a pale blue effect.

' That which is up at my window has been out in
the air four hours, and it, when wetted, comes out a
strong blue tint, as No. 6 of the scale. The day is dry
but with no sun, the lower region pretty clear but
clouds above.

' After reading your notes and examining the illustrations, I could not resist writing to you, though, as
you see, I have nothing to say.

' Ever truly yours,
' M. FARADAY.'

II.

This year he was made Corresponding Associate of
the Accademia Pontifica, Rome, and Foreign Associate
of the Academy of Sciences, Haarlem.

An interesting letter from M. Quetelet, and one of
Faraday's to M. Becquerel, show the position which he
held in science.

M. QUETELET TO FARADAY.

' Bruxelles, le 9 septembre 1850.

' Monsieur et très-illustre confrère,—Je me suis aperçu
avec un regret infini que vous n'avez pas encore reçu
le diplôme d'associé de notre académie. M. le docteur
Pincoffs, qui passe par Londres, a bien voulu se charger
de réparer cette négligence de la personne qui aurait
dû vous faire parvenir ce diplôme depuis longtemps.
J'y joins deux volumes des dernières publications de
notre académie, qui vous sont destinés.

'Je suis très-heureux de trouver cette occasion pour vous exprimer mes sentiments de haute estime et ma reconnaissance pour toutes les bontés que vous ne cessez de me témoigner. Vos encouragements ont été pour moi les plus précieuses récompenses que je pusse ambitionner, et vous me les avez toujours prodigués avec la plus grande bienveillance.

'Vous m'avez montré un recueil très-précieux de portraits de personnes auxquelles vous portez de l'affection. Il y aurait, peut-être, de la fatuité à vous demander une place auprès d'elles ; si, cependant, vous n'y voyez pas trop d'outrecuidance, permettez-moi de vous offrir un exemplaire d'un de mes portraits, dessiné par mon beau-frère. Je serais heureux surtout s'il pouvait me valoir en retour le portrait d'un des hommes dont j'estime le plus le talent et le noble caractère.

'Agréez, je vous prie, mon cher et illustre confrère, mes compliments les plus distingués et les plus affectueux.

'QUETELET.'

To M. Becquerel, who wrote to him regarding the magnetism of oxygen, he replied :—

FARADAY TO M. BECQUEREL.

'Royal Institution : December 30, 1850.

'My dear Sir,—It is with great pleasure that I receive a letter from you, for much as I have thought of your name and the high scientific labours connected with it, I do not remember that I have seen your handwriting before. I shall treasure the letter in a certain volume of portraits and letters that I keep devoted to the personal remembrance of the eminent men who adorn

science whom I have more or less the honour and delight
of being acquainted with.

'In reference to the queries in your letter, I suppose the following will be sufficient answer.

'I developed and *published* the nature and principles of the action of magnetic and diamagnetic media upon substances *in* them, more or less magnetic or diamagnetic than themselves, in the year 1845, or just *five* years ago. The paper was read at the Royal Society, January 8, 1846, and is contained in the "Philosophical Transactions" for 1846, p. 50, &c. If you refer to the numbered paragraphs 2357, 2363, 2367, 2400k, 2406, 2414, 2423, 2438, you will see at once how far I had gone at that date. The papers were republished in Poggendorf's "Annalen," and I believe in the Geneva, the Italian, and German journals in one form or another.

'In reference to the magnetism of oxygen, *three* years ago, i.e. in 1847, I showed its high magnetic character in relation to nitrogen and all other gases, and that air owed its place amongst them to the oxygen it contained. I even endeavoured to analyse the air, separating its oxygen and nitrogen by magnetic force, for I thought such a result possible. All this you will find in a paper published in the "Philosophical Magazine," for 1847, vol. xxxi. page 401, &c. This paper was also published at full length in Poggendorff's "Annalen," 1848, vol. lxxiii. page 256, &c. I shall send you a copy of it immediately by M. Bailliere, who has undertaken to forward it to you. I have marked it in ink to direct your attention. In it also you will find the effect of *heat on oxygen and air*; the experiments were all devised and made upon the principles before

developed, concerning the mutual relation of substances
and the media surrounding them.

'This year I have been busy extending the above
researches, and have sent in several papers to the Royal
Society, and have also given a Bakerian lecture in which
they were briefly summed up. I fortunately have a copy
in slips of the Royal Society's abstract of these papers,
and therefore will send it with the paper from the " Philosophical Magazine." I suppose it will appear in the
outcoming number of the " Philosophical Magazine."
The papers themselves are now in the hands of the
printer of the " Transactions."

'I was not aware, until lately, of that paper of M.
Edmond Becquerel, to which you first refer. My health
and occupation often prevent me from reading up to
the present state of science. Immediately that I knew
of it, I added a note (by permission) to my last paper,
series xxvi., in which I referred to it, and quoted at
length what it said in reference to atmospheric magnetism, calling attention also to my own results as to
oxygen three years ago, and those respecting media
five years ago. I have no copy of this note, or
I would send it to you. It was manifest to me that
M. Edmond Becquerel had never heard of my results,
and though that makes no excuse to myself, I hope it
will be to him a palliation that I had not before heard
of his. The second one I had not heard of until I
received your letter the day before yesterday.

'I was exceedingly struck with the beauty of M.
E. Becquerel's experiments, and though the differential
balance I have described in my last paper will, I expect,
give me far more delicate indications, when the perfect
one, which is in hand, is completed, still I cannot

express too freely my praise of the apparatus and
results which the first paper describes and which is
probably surpassed by those in the second.

'I know the severe choice of your Academy of
Sciences, and I also know that France has ever been
productive of men who deserve to stand as candidates,
whenever a vacancy occurs in any branch of knowledge ;
and though, as you perceive, I do not know all that M. E.
Becquerel has done, I know enough to convince me that he
deserves the honour of standing in that body and to create
in me strong hopes that he will obtain his place there.

'Ever, my dear M. Becquerel, your faithful admirer,

'M. FARADAY.'

III.

Letters to Miss Martineau, Dr. du Bois Raymond,
Dr. Tyndall, and Professor Oersted, show his truthfulness, his kindness, and his thoughtfulness in his ordinary
correspondence, by which he gained always the goodwill and often the affection of those who were personally unknown to him.

FARADAY TO DR. DU BOIS RAYMOND.

'Royal Institution : January 15, 1850.

'Dear Sir,—I this day received your kind present of
books—your great work—and also the letter. I regret
that I have no better thanks to offer you than those of
a man who cannot estimate the work properly. I look
with regret at the pages, which are to me a sealed
book ; and, but that increasing infirmities too often
warn me off, I would even now attack the language
of science and knowledge, for such the German language is.

'Mr. Magnus, whom I rejoice to call a friend, told
me of your great experiment, in which, from the muscular excitement of the living human being, you obtained
a current of electricity.   I endeavoured, a few months
ago, to procure the result, but did not succeed.   No
doubt, being unacquainted with all the precautions
needful, and the exact manner of proceeding, I was at
fault.   And now I am so engaged by the duties of my
station and the season, that I have no time for anything else.   During the season I trust to pick up the
information that will give me success the next time
that I try.

'The second copy of your work is already on the
road to the Royal Society, and I shall do all I can to
direct the attention of men of science and others to the
copy you have sent me, by placing it before them on
the tables of this Institution.

'I am, Sir, your very obliged and grateful servant,
                                         'M. FARADAY.'

FARADAY TO PROFESSOR OERSTED.

'Royal Institution: March 15, 1850.

'My dear Sir,—I received your very kind letter
two or three weeks ago, and was very greatly gratified
that you should remember me.

.   .   .   .   .   .

'This is a time of the year in which formal matters
occupy me so much, that (together with a system soon
wearied) they prevent me from working to any good
purpose ; so that I have little or nothing to say.

'I have, it is true, sent a paper to the Royal Society,[1]

---

[1] Twenty-third series, on polar or other condition of diamagnetic bodies.

two or three months ago, which was read lately ; and
in it I describe my failure to produce the results of
Weber, Reich, and some others, or (of such as were
produced) my reference of them to other principles of
action than those they had adopted.   This branch of
science is in a very active and promising state.   Many
men—and amongst them yourself—are working at it,
and it is not wonderful that views differ at first.

' Time will gradually sift and shape them, and I
believe that we have little idea at present of the importance they may have ten or twenty years hence.

' As soon as my paper is printed I shall send it to
you, and, I hope, with copies of those you have not
received.   I thought I had sent you all in order ; for
it was to me a delight to think I might do so.   I do
not know what can have come in the way of them ; but
if I have copies left, you shall have them with the
next paper.

.   .   .   .   .   .

' I am, my dear Sir, your very obliged and faithful
servant,

'M. FARADAY.'

He writes the following letter regarding the production of the Acarus Crossii by electricity.

FARADAY TO MISS MARTINEAU.

' Royal Institution : April 11, 1850.

' My dear Madam,—I am sorry to find that in your
great work you have been led, at p. 417, vol. ii., into
an error respecting me by an authority which you
might well think sufficient, but which is inaccurate.   I
cannot understand how the error arose at first ; but it

appeared in the papers, and I found it necessary, in a letter to the Editor of the " Literary Gazette," March 4, 1837, page 147, to correct it. The error probably passed from the papers into the " Annual Register," and from that into the far more important position it holds in your History.

.   .   .   .

'I hope you will forgive me for writing to you about this matter. I feel it a great honour to be borne on your remembrance, but I would not willingly be there in an erroneous point of view.

'I have the honour to be, my dear Madam, with every respect, your faithful, humble servant,

'M. FARADAY.'

Miss Martineau replied :

'April 13.

'I am greatly obliged to you for correcting the mistake in my History regarding your countenance of the Acarus Crossii. . . .

'It never occurred to me to doubt the authority of the " Annual Register" in a matter of such straight-forward contemporary statement, and it is really difficult to see how one can make sure of one's material. . . .

'Believe me, my dear Sir, with the highest respect, your obliged,

'HARRIET MARTINEAU.'

FARADAY TO JOHN TYNDALL, ESQ.

'Royal Institution : November 19, 1850.

'Dear Sir,—I do not know whether this letter will find you at Marburg, but, though at the risk of missing you, I cannot refrain from thanking you for your

kindness in sending me the rhomboid of calcareous
spar. I am not at present able to pursue that subject, for
I am deeply engaged in terrestrial magnetism ; but I
hope some day to take up the point respecting the
magnetic condition of associated particles.   In the
meantime, I rejoice at every addition to the facts and
to the reasoning connected with the subject.   It is
wonderful how much good results from different persons
working at the same matter.   Each one gives views
and ideas new to the rest.   When science is a republic,
then it gains ; and though I am no republican in other
matters, I am in that.

' With many thanks for your kindness, I am, Sir,
your very obedient servant,

'   ' M. FARADAY.'

I.

In 1851, in July, September, October, November,
and December, much experimental work was done in
the laboratory.   The results were sent to the Royal
Society in the twenty-eighth series of ' Researches ' on
lines of magnetic force, their definite character, and
their distribution in a magnet and throughout space ;
and in the twenty-ninth series on the employment of
the induced magneto-electric current as a test and
measure of magnetic forces.

Dr. Tyndall's account of the relation of these papers
to Faraday's speculations on the nature of matter and
lines of force is of the highest interest, as it sets forth
clearly the working of Faraday's mind.

Dr. Tyndall says : ' The scientific picture of Faraday
would not be complete without a reference to his
speculative writings.   On Friday, January 19, 1844,

he opened the weekly evening-meetings of the Royal
Institution by a discourse entitled " A speculation touching Electric Conduction and the nature of Matter." In
this discourse he not only attempts the overthrow of
Dalton's theory of atoms, but also the subversion of
all ordinary scientific ideas regarding the nature and
relations of matter and force.   He objected to the use
of the term atom :—" I have not yet found a mind,"
he says, " that did habitually separate it from its accompanying temptations; and there can be no doubt
that the words definite proportions, equivalent, primes,
&c., which did and do fully express all the *facts* of
what is usually called the atomic theory in chemistry,
were dismissed because they were not expressive
enough, and did not say all that was in the mind
of him who used the word atom in their stead."

    ' He then tosses the atomic theory from horn to horn
of his dilemmas.   What do we know, he asks, of the
atom apart from its force?   You imagine a nucleus
which may be called *a*, and surround it by forces which
may be called *m*; " to my mind the *a* or nucleus
vanishes, and the substance consists in the powers of
m.   And, indeed, what notion can we form of the
nucleus independent of its powers?   What thought
remains on which to hang the imagination of an *a*
independent of the acknowledged forces? "   Like Boscovich, he abolishes the atom and puts a " centre of force "
in its place.

    ' With his usual courage and sincerity, he pushes his
view to its utmost consequences.   " This view of the
constitution of matter," he continues, " would seem to
involve necessarily the conclusion that matter fills all
space, or at least all space to which gravitation extends;

for gravitation is a property of matter dependent on a   1851.
certain force, and it is this force which constitutes the   Æt.59–60.
matter. In that view, matter is not merely mutually
penetrable;[1] but each atom extends, so to say, throughout the whole of the solar system, yet always retaining
its own centre of force."

'It is the operation of a mind filled with thoughts
of this profound, strange, and subtle character that we
have to take into account in dealing with Faraday's
later researches. A similar cast of thought pervades a
letter addressed by Faraday to Mr. Richard Phillips,
and published in the "Philosophical Magazine" for
May, 1846. It is entitled "Thoughts on Ray-vibra-
tions," and it contains one of the most singular speculations that ever emanated from a scientific mind. It
must be remembered here, that though Faraday lived
amid such speculations, he did not rate them highly,
and that he was prepared at any moment to change
them or let them go. They spurred him on, but they
did not hamper him. His theoretic notions were *fluent*;
and when minds less plastic than his own attempted to
render those fluxional images rigid, he rebelled. He
warns Phillips, moreover, that from first to last " he
merely threw out as matter for speculation the vague
impressions of his mind ; for he gave nothing as the
result of sufficient consideration, or as the settled conviction, or even probable conclusion at which he had
arrived."

'The gist of this communication is that gravitating
force acts in lines across space, and that the vibrations of

---

[1] He compares the interpenetration of two atoms to the coalescence of
two distinct waves, which, though for a moment blended to a single
mass, preserve their individuality, and afterwards separate.

light and radiant heat consist in the tremors of these lines of force. "This notion," he says, "as far as it is admitted, will dispense with the ether, which, in another view, is supposed to be the medium in which these vibrations take place." And he adds, further on, that his view " endeavours to dismiss the ether, but not the vibrations." The idea here set forth is the natural supplement of his previous notion that it is gravitating force which constitutes matter, each atom extending, so to say, throughout the whole of the solar system.

'The letter to Mr. Phillips winds up with this beautiful conclusion :—"I think it likely that I have made many mistakes in the preceding pages, for even to myself my ideas on this point appear only as the shadow of a speculation, or as one of those impressions upon the mind which are allowable for a time as guides to thought and research. He who labours in experimental inquiries, knows how numerous these are, and how often their apparent fitness and beauty vanish before the progress and development of real natural truth."

'Let it then be remembered that Faraday entertained notions regarding matter and force altogether distinct from the views generally held by scientific men. Force seemed to him an entity dwelling along the line in which it is exerted. The lines along which gravity acts between the sun and earth seem figured in his mind as so many elastic strings : indeed, he accepts the assumed instantaneity of gravity as the expression of the enormous elasticity of the " lines of weight." Such views, fruitful in the case of magnetism, barren as yet in the case of gravity, explain his efforts to transform this latter force. When he goes into the open air and permits his helices to fall, to his mind's eye they are tearing through the lines of gravitating power, and hence his hope and conviction that an effect would and ought to be produced. It must ever be borne in mind that Faraday's difficulty in dealing with these conceptions was at bottom the same as that of Newton ; that he is, in fact, trying to overleap this difficulty, and with it probably the limits prescribed to the intellect itself.

1851.

Æт.59–60.

' The idea of lines of magnetic force was suggested to Faraday by the linear arrangement of iron filings when scattered over a magnet. He speaks of, and illustrates by sketches, the deflection, both convergent and divergent, of the lines of force, when they pass respectively through magnetic and diamagnetic bodies. These notions of concentration and divergence are also based on the direct observation of his filings. So long did he brood upon these lines ; so habitually did he associate them with his experiments on induced currents, that the association became " indissoluble," and he could not think without them. " I have been so accustomed," he writes, " to employ them, and especially in my last researches, that I may have unwittingly become prejudiced in their favour, and ceased to be a clear-sighted judge. Still, I have always endeavoured to make experiment the test and controller of theory and opinion ; but neither by that nor by close cross-examination in principle, have I been made aware of any error involved in their use."

' In his later researches on magne-crystallic action, the idea of lines of force is extensively employed ; it, indeed, led him to an experiment which lies at the root of the whole question. In his subsequent researches on atmospheric magnetism, the idea receives still wider application, showing itself to be wonderfully flexible and convenient. Indeed, without this conception, the attempt to seize upon the magnetic actions, possible or actual, of the atmosphere would be difficult in the extreme ; but the notion of lines of force, and of their divergence and convergence, guides Faraday without perplexity through all the intricacies of the question. After the completion of those researches, and in a paper forwarded to the Royal Society on October 22, 1851, he devotes himself to the formal development and illustration of his favourite idea. The paper bears the title " On Lines of Magnetic Force, their definite character, and their distribution within a magnet and through space." A deep reflectiveness is the characteristic of this memoir. In his experiments, which are perfectly beautiful and profoundly suggestive, he takes but a secondary delight. His object is to illustrate the utility of his conception of lines of force. " The study of these lines," he says, " has at different times been greatly influential in leading me to various results which I think prove their utility as well as fertility."

' Faraday for a long period used the lines of force merely as " a representative idea." He seemed for a time averse to going further in expression than the lines themselves, however much further he may have gone in idea. That he believed them to exist at all times round a magnet, and irrespective of the existence of magnetic matter, such as iron-filings, external to the magnet, is certain. No doubt the space round every magnet presented itself to his imagination as traversed by loops of magnetic power, but he was chary in speaking of the physical substratum of those loops. Indeed, it may be doubted whether the *physical theory* of lines of force presented itself with any distinctness to his own mind. The possible complicity of the luminiferous ether in magnetic phenomena was certainly in his thoughts. " How the magnetic force," he writes, " is transferred through bodies or through space we know not: whether the result is merely action at a distance, as in the case of gravity; or by some intermediate agency, as in the case of light, heat, the electric current, and (as I believe) static electric action. The idea of magnetic fluids, as applied by some, or of magnetic centres of action, does not include that of the latter kind of transmission, *but the idea of lines of force does.*" And he continues thus :—" I am more inclined to the notion that in the transmission of the [magnetic] force there is such an action [an intermediate agency] external to the magnet, than that the effects are merely attraction and repulsion at a distance. *Such an affection may be a function of the ether: for it is not at all unlikely that, if there be an ether, it should have other uses than simply the conveyance of radiations.*" When he speaks of the magnet in certain cases " revolving amongst its own forces," he appears to have some conception of this kind in view.

'A great part of the investigation completed in October, 1851, was taken up with the motions of wires round the poles of a magnet, and the converse. He carried an insulated wire along the axis of a bar magnet from its pole to its equator, where it issued from the magnet, and was bent up so as to connect its two ends. A complete circuit, no part of which was in contact with the magnet, was thus obtained. He found that when the magnet and the external wire were rotated together no current was produced; whereas, when *either* of them was rotated, and the other left at rest, currents were evolved. He then abandoned the axial wire, and allowed the magnet itself to take its place; the result was the same.[1] It was the *relative* motion of the magnet and the loop that was effectual in producing a current.

'The lines of force have their roots in the magnet, and though they may expand into infinite space, they eventually return to the magnet. Now, these lines may be intersected close to the magnet or at a distance from it. Faraday finds *distance* to be perfectly immaterial so long as the *number* of lines intersected is the same. For example, when the loop connecting the equator and the pole of his bar-magnet performs one complete revolution round the magnet, it is manifest that all the lines of force issuing from the magnet are *once* intersected. Now, it matters not whether the loop be ten feet or ten inches in length, it matters not how it may be twisted and contorted, it matters not how near to the magnet, or how distant from it the loop may be, one revolution always produces the same amount of current electricity, because in all these cases all the lines of force issuing from the magnet are *once* intersected and no more.

'From the external portion of the circuit he passes in idea to the internal, and follows the lines of force into the body of the magnet itself. His conclusion is that there exist lines of force within the magnet of the same *nature* as those without. What is more, they are exactly equal in *amount* to those without. They have a relation in *direction* to those without; and, in

[1] In this form the experiment is identical with one made twenty years earlier.

fact, are continuations of them. . . . "Every line of

force, therefore, at whatever distance it may be taken from the magnet, must be considered as a closed circuit, passing in some part of its course through the magnet, and having an equal amount of force in every part of its course."

'All the results here described were obtained with *moving metals.* "But," he continues with profound sagacity, "mere motion would not generate a relation, which had not a foundation in the existence of some previous state ; and therefore the *quiescent* metals must be in some relation to the active centre of force," that is, to the magnet. He here touches the core of the whole question, and when we can state the condition into which the conducting wire is thrown *before* it is moved, we shall then be in a position to understand the physical constitution of the electric current generated by its motion.

'In this inquiry Faraday worked with steel magnets, the force of which varies with the distance from the magnet. He then sought a *uniform field* of magnetic force, and found it in space as affected by the magnetism of the earth. His next memoir, sent to the Royal Society on December 31, 1851, is "On the Employment of the Induced Magneto-electro Current as a Test and Measure of Magnetic Forces." He forms rectangles and rings, and by ingenious and simple devices collects the opposed currents which are developed in them by rotation across the terrestrial lines of magnetic force. He varies the shapes of the rectangles while preserving their areas constant, and finds that the constant area produces always the same amount of current per revolution. The current depends solely on the number of lines of force intersected, and when this number is kept constant the current remains constant too. Thus the lines of magnetic force are continually before his eyes ; by their aid he colligates his facts, and through the inspirations derived from them he vastly expands the boundaries of our experimental knowledge. The beauty and exactitude of the results of this investigation are extraordinary. I cannot help thinking, while I dwell upon them, that the discovery of magneto-electricity is the greatest experimental result ever obtained by an investigator. It is the Mont Blanc of Faraday's own achievements. He always worked at great elevations, but a higher than this he never subsequently attained.'

For the Institution, after Easter, he gave six lectures on some points of electrical philosophy.

He ends the notes of this course thus : ' The *truth* of science has ever had not merely the task of evolving herself from the dull uniform mist of ignorance ; but also that of repressing and dissolving the phantoms of imagination which ever rise up in new and tempting shapes, and which, not being of her, crowd before and around her, and embarrass her in her way.'

He gave three Friday discourses :—

On the magnetic characters and relation of oxygen and nitrogen ; on atmospheric magnetism (in this lecture he stated that the magnetism of the earth equalled 8464,000,000,000,000,000,000 one pound magnet bars) ; and on Schönbein's ozone.

He gave the Christmas Lectures on the forces of matter.

The letters which he wrote to Professor Schönbein show his interest in ozone, and his delight in making the discoveries of his friend known in England.

FARADAY TO PROFESSOR SCHÖNBEIN.

'Royal Institution: March 5, 1851.

'My dear Friend,—I had your hearty Christmas letter in due time.

.   .   .   .   .   .

'My, or rather your, evening, I expect will be June 13, or the middle of our Great Exhibition. When I drew out a sort of preliminary sketch of the subject, I was astonished at the quantity of matter—real matter—and its various ramifications; and it seems still to grow upon me. What you will make it before I begin to talk, I do not know.

'I do not as yet see any relation between the magnetic condition of oxygen and the ozone condition, but who can say what may turn up? I think you make an inquiry or two as to the amount of magnetic force which oxygen carries into its compounds. This is indeed a wonderful part of the story, for magnetic as *gaseous oxygen* is, the substance seems to lose all such force in compounds. Thus water, which is $\frac{8}{9}$ths oxygen, contains no sensible trace of it; and peroxide of iron, which itself consists of two most magnetic constituents, is scarcely sensibly magnetic; so little have either of these bodies carried their forces into the resulting compound.

'Sometimes I think we may understand a little better such changes by thinking that magnetism is a physical rather than a chemical force, but after all such a difference is a mere play upon words, and shows ignorance rather than understanding. But you know there are really a great many things we are as yet ignorant of,

and amongst the rest the infinitesimal proportion of
our knowledge to that which really *is to be known.*

.   .   .   .   .   .

' I have no doubt I answer your letters very badly,
but, my dear friend, *do you remember* that *I forget,* and
that I can no more help it than a sieve can help the
water running out of it.   Still you know me to be your
old and obliged and affectionate friend, and all I can say
is, the longer I know you the more I seem to cling to
you.   Ever, my dear Schönbein, yours affectionately,

' M. FARADAY.'

FARADAY TO PROFESSOR SCHÖNBEIN.

'Hastings : April 19, 1851.

' My dear Schönbein,—Here we are at the seaside ;
and my mind so vacant (not willingly) that I cannot
get an idea into it.   You will wonder, therefore, why I
write to you, since I have nothing to say, but the fact
is I feel as if I owed you a letter, and yet cannot
remember clearly how that is.   Still I would rather
appear stupid to you than oblivious of your kindness,
and yet very forgetful I am.   In six or seven weeks I
shall be talking of ozone.   I hope I shall not discredit
you, or fail in using well all the matter you have given
me, abundant and beautiful as it is.   But I feel that my
memory does not hold things together in hand as it
used to do.   Formerly I did not care about the multi-
plicity of items ; they all took their place, and I picked
out what I wanted at pleasure.   Now I am conscious
of but few at once, and it often happens that a feeble
point which has present possession of the mind obscures
from recollection a stronger and better one which is

ready and waiting.   But we must just do the best we
can, and you may be sure I will do as well for you as
I would for myself.

' I set about explaining the other evening my views of
atmospheric magnetism, and found when I had done
that I had left out the two or three chief points.   I
only hope the printed papers contain them, and that
they will be found good by the men who are able to
judge.

.   .   .   .   .   .

' I am, as ever, most truly yours,

' M. FARADAY.'

FARADAY TO PROFESSOR SCHÖNBEIN.

'Tynemouth : August 1, 1851.

' My dear Schönbein,—. . . The ozone evening went
off wonderfully well ; our room overflowed, and many
went away unable to hear (my account at least) of this
most interesting body.   Through your kindness, the
matter was most abundant and instructive, and the
experiments very successful.   The subject has been sent
into the world so much piecemeal, that many were astonished to see how great it became when it was presented
as one whole, and yet my whole must have been a most
imperfect sketch, for I found myself obliged to abridge
my thoughts in every direction.   Many accounts were
printed by different parties, and some very inaccurately,
since they had to catch up what they could.   A notice of
four pages appeared in the " Proceedings " of the Royal
Institution.

.   .   .   .   .   .

· The subject excited great interest, and, from what

the folks said, I had no reason to be ashamed either for the subject or myself.

'And now, my dear Schönbein, I am very weary. Perhaps to-day you are at Glarus. I was two days at Ipswich at our meeting—no more, for want of strength. Queen's balls, Paris fêtes, &c. &c., I am obliged (and very willingly) to leave all to others.

'I am, ever yours,
'M. FARADAY.'

A letter to General Portlock, who was then deputy governor of the Royal Military Academy, Woolwich, gives some insight into his views on the way in which chemistry should be taught.

FARADAY TO GENERAL PORTLOCK.

'December 1, 1851.

'My dear Portlock,—. . . As one of the Senate of the University of London, and appointed with others especially to consider the best method of examination, I have had to think very deeply on the subject, and have had my attention drawn to the practical working of different methods at our English and other Universities; and know there are great difficulties in them all. Our conclusion is that examination by papers is the best, accompanied by *viva voce* when the written answers require it. Such examinations require that the students should be collected together, each with his paper, pens, and ink; that each should have the paper of questions (before unknown) delivered to him; that they should be allowed three, or any sufficient number of hours to answer them, and that they should be carefully watched by the examiner or some other officer, so as to prevent their having any communication with each other, or going out of the room for that time. After which, their written answers have to be taken and examined carefully by the examiner and decided upon according to their respective merits. We think that no numerical value can be attached to the questions, because everything depends on how they are answered; and that is the reason why I am not able to send you such a list at the present time.

'My verbal examinations at the Academy go for very little, and were instituted by me mainly to keep the students' attention to the lecture for the time, under the pressure of a thought that inquiry would come at the end. My instructions always have been to look to the note-books for the result; and so the verbal examinations are only used at last as confirmations or corrections of the conclusions drawn from the notes.

'I should like to have had a serious talk with you on this matter, but my time is so engaged that I cannot come to you at Woolwich for the next two or three weeks, so I will just jot down a remark or two. In the first place, the cadets have only the lectures, and no practical instruction in chemistry, and yet chemistry is eminently a practical science. Lectures alone cannot be expected to give more than a general idea of this most extensive branch of science, and it would be too much to expect that young men who at the utmost hear only fifty lectures on chemistry, should be able to answer with much effect in writing, to questions set down on paper, when we know by experience that daily work for eight hours in *practical laboratories* for *three months* does not go very far to confer such ability.

'Again: the audience in the lecture-room at the Academy always, with me, consists of four classes, i.e. persons who have entered at such different periods as to be in four different stages of progress. It would, I think, be unfair to examine all these as if upon the same level; they constitute four different classes, and we found it in our inquiries most essential to avoid mixing up a junior and a senior class one with the other. Even though it were supposed that you admitted only those who were going out to examination, and such others from the rest as chose to volunteer, yet as respects them it has to be considered that I may not go on from the beginning to the end of their fifty lectures increasing the importance and weight of the matter brought before them, for I have to divide the fifty into two courses, each to be begun and finished in the year, and I ever have to keep my language and statements so simple as to be fit for mere beginners and not for advanced pupils.

'I have often considered whether some better method of giving instruction in chemistry to the cadets could not be devised, but have understood that it was subordinate to other more important studies, and that the time required by a practical school, which is considerable, could not be spared. Perhaps, however, you may have some view in this direction, and I hasten to state to you what I could more earnestly and better state by word of mouth, that you must not think me the least in the way. I should be very happy, by consultation, in the first instance, to help you in such a matter, though I could not undertake any part in it. I am getting older, and find the Woolwich duty, taking in as it does large parts of two days, as much as I can manage with

satisfaction to myself; so that I could not even add on
to it such an examination by written papers as I have
talked about: but I should rejoice to know that the
whole matter was in more practical and better hands.

'Ever, my dear Portlock, yours very truly,

'M. FARADAY.

'I refused to be an examiner in our University.

'M. F.'

In 1851 no work was recorded for the Trinity House.

II.

During this year he was made Member of the Royal
Academy of Sciences at the Hague, Corresponding
Member of the Batavian Society of Experimental
Philosophy, Rotterdam; Fellow of the Royal Society
of Sciences, Upsala; a Juror of the Great Exhibition.

III.

The freedom of his mind from jealousy of those who
were at work on his subjects, is again seen in his
letters to M. Becquerel and to Dr. Tyndall; and in a
later letter to Dr. Tyndall he shows how thoughtfully
he could give advice when asked for it.

FARADAY TO M. BECQUEREL.

'Royal Institution: January 17, 1851.

'My dear M. Becquerel,—I received your letter of
the 14th instant yesterday, and hasten to reply to it,
as you desire; first, however, thanking you for your kind
expressions, which will be a strong stimulus to me,

coming as they do from a master in science. I would not have you for a moment think that I put my paper of three years ago and that of M. E. Becquerel's of last year on the same footing, except in this, that we each discovered for ourselves at those periods the high magnetic relation of oxygen to the other gases. M. E. Becquerel has made excellent measurements, which I had not, and his paper is, in my opinion, a most important contribution to science.

'I am not quite sure whether you are aware that in my paper of 1847 the comparison of one gas with another is always at the *same* temperature, i.e. at common temperatures, and it was a very striking fact to me to find that oxygen was magnetic in relation to hydrogen to such an extent as to be equal in attractive force to its force of gravity, for the oxygen was suspended in the hydrogen by magnetic force alone, " Phil. Mag." xxxi. pp. 415, 416. I do not think that much turns upon the circumstance of calling oxygen magnetic or diamagnetic in 1847, when the object was to show how far oxygen was apart from the other gases in the magnetic direction, these terms being employed in relation to other bodies, and with an acknowledgment that the place of zero was not determined. If I understand rightly, M. Edmond Becquerel still calls bismuth and phosphorus magnetic, whilst I call them diamagnetic. He considers space as magnetic: I consider it as zero. If a body should be found as eminently diamagnetic in my view as iron is magnetic, still I conclude M. Edmond Becquerel would consider it magnetic. He has not yet adopted the view of any zero or natural standard point. But this does not prevent us from fully understanding each other, and the facts upon which the

distinction of oxygen from nitrogen and other gases   1851.
are founded, remain the same, and are just as well made Ær.59-60.
known by the one form of expression as the other.
It was, therefore, to me a great delight, when I first saw
his paper in last November, to have my old results
confirmed and so beautifully enlarged in the case of
oxygen and nitrogen by the researches of M. E. Becquerel,
and beyond all to see the beautiful system of measure-
ment applied to them which is described in his published paper.

' Pray present my kindest remembrances and wishes
to him, and believe me to be, with the highest
respect, my dear M. Becquerel, your faithful, obliged
servant,

'M. FARADAY.'

FARADAY TO DR. J. TYNDALL.

'Hastings: April 19, 1851.

' Dear Sir,—Whilst here, resting for a while, I take
the opportunity of thanking you for your letter of
February 4, and also for the copy of the paper in the
" Philosophical Magazine," which I have received. I
had read the paper before, and was very glad to have
the development of your researches more at large than
in your letter. Such papers as yours make me feel
more than ever the loss of memory I have sustained,
for there is no reading them, or at least retaining the
argument, under such deficiency.

' Mathematical formulæ more than anything require
quickness and surety in receiving and retaining the
true value of the symbols used, and when one has to

look back at every moment to the beginning of a paper, to see what H or A or B mean, there is no making way. Still, though I cannot hold the whole train of reasoning in my mind at once, I am able fully to appreciate the value of the results you arrive at, and it appears to me that they are exceedingly well established and of very great consequence. These elementary laws of action are of so much consequence in the development of the nature of a power which, like magnetism, is as yet new to us.

'My views with regard to the cause of the annual, diurnal, and some other variations are not yet published, though printed. The next part of the "Philosophical Transactions" will contain them. I am very sorry I am not able to send you a copy from those allowed to me, but I have had so many applications from those who had some degree of right that they are all gone. I only hope that when you see the "Transactions" you may find reason to think favourably of my hypothesis. Time does not lessen my confidence in the view I have taken, but I trust when relieved from my present duties, and somewhat stronger in health, to add experimental results regarding oxygen, so that the mathematician may be able to take it up.

'As you say, in the close of your letter, I have far more confidence in the one man who works mentally and bodily at a matter, than in the six who merely talk about it, and I therefore hope and am fully persuaded that you are working.

'Nature is our kindest friend and best critic in experimental science, if we only allow her intimations to fall unbiassed on our minds. Nothing is so good as an experiment which, whilst it sets an error right, gives (as a reward for our humility in being reproved) an absolute advancement in knowledge.

'I am, my dear Sir, your very obliged and faithful servant,   'M. FARADAY.'

FARADAY TO DR. TYNDALL.

'Tynemouth : August 1, 1851.

'My dear Sir,—. . . . In the first place, many thanks for the specimens which I shall find presently at home. I was very sorry not to see you make your experiments, but hope to realise the pressure results, which interest me exceedingly.   I want to have a very clear view of them.

'But now for the Toronto matter.   In such a case, private relationships have much to do in deciding the matter ; but if you are comparatively free from such considerations, and have simply to balance your present power of doing good with that you might have at Toronto, then I think I should (in your place) choose the latter.   I do not know much of the University, but I trust it is a place where a man of science and a true philosopher is required, and where, in return, such a man would be nourished and cherished in proportion to his desire to advance natural knowledge.   I cannot doubt, indeed, that the University would desire the advancement of its pupils, and also of knowledge itself, so I think that you would be exceedingly fit for the position, and I hope the position fit for you.   If I had any power of choosing or recommending, I should aid your introduction into the place, both because I know what you have already done for science, and I heard[1] how you

---

[1] At the British Association at Ipswich.

could state your facts and touch your audience. Now
I do not, for I cannot, proffer you a certificate, because
I have in every case refused for many years past to give
any on the application of candidates. Neither, indeed,
have you asked me for one. Nevertheless, I wish to
say that when I am asked about a candidate by those
who have the choice or appointment, I never refuse to
answer; and, indeed, if my opinion would be useful,
and there was a need for it, you might use this letter
as a private letter, showing it, or any part of it, to
any whom it might concern.

'And now you must excuse me from writing any
more, for my muscles are stiff and weak, and my head
giddy.

'Ever, my dear Dr. Tyndall, yours most truly,

'M. Faraday.'

I.

In 1852 and 1853 there was not much original
research. In the autumn of 1852 he worked hard on
magnetic force; on December 9 he wrote to Professor
Schönbein.

FARADAY TO PROFESSOR SCHÖNBEIN.

'Brighton.

'My dear Friend,—If I do not write to you now
I do not know when I shall; and if I write to you now
I do not know what I shall say, for I am here sleeping,
eating, and lying fallow, that I may have sufficient
energy to give half a dozen Juvenile Christmas Lectures. The fact is, I have been working very hard for
a long time to no satisfactory end; all the answers I

have obtained from nature have been in the negative, 1852–53. and though they show the truth of nature as much as Æt.60–62. affirmative answers, yet they are not so encouraging, and so for the present I am quite worn out. I wish I possessed some of your points of character; I will not say which, for I do not know where the list might end, and you might think me simply absurd, and, besides that, ungrateful to Providence.

.   .   .   .

'Your letter quite excites me, and I trust you will establish undeniably your point. It would be a great thing to trace the state of combined oxygen by the colour of its compound, not only because it would show that the oxygen had a special state, which could in the compound produce a special result, but also because it would, as you say, make the optical effect come within the category of scientific appliances, and serve the purpose of a philosophic indication and means of research, whereas it is now simply a thing to be looked at. Believing that there is nothing superfluous or deficient, or accidental, or indifferent in nature, I agree with you in believing that colour is essentially connected with the physical condition and nature of the body possessing it; and you will be doing a very great service to philosophy, if you give us a hint, however small it may seem at first, in the development, or, as I may even say, in the perception of this connection.'

For the Institution in 1852, after Easter, he gave a course of six lectures on points connected with the non-metallic elements: he took oxygen, chlorine, hydrogen, nitrogen, sulphur and carbon.

In his first lecture, on oxygen, he says : ' It can rest in the state of combination, half way on towards its final end, as it were,—gun-cotton, gun-sawdust. These cases indicate something of the mystery of combination, and the probable work of oxygen in more recondite cases, as in the living system.   Whilst in health all advances, and well ; but even a life is but a chemical act prolonged. If death occur, then more rapidly, oxygen and the affinities run on to the final state.'

The first and last Friday discourses were given on lines of magnetic force, and at the first meeting of the members in 1853, he gave a lecture which contained the result of his work during the previous autumn ; the title was ' Observations on Magnetic Force.' For the more careful study of the magnetic power, he had a torsion balance constructed of a peculiar kind, and with it he tried to investigate the right application of the law of the inverse square of the distance, as the universal law of magnetic action. By the end of his lecture, he showed that he was not satisfied with his results. He says : ' Before leaving this first account of recent experimental researches, it may be as well to state, that they are felt to be imperfect, and may perhaps even be overturned ; but that as such a result is not greatly anticipated, it was thought well to present them to the members of the Royal Institution and the scientific world, if peradventure they might excite criticism and experimental examination, and so aid in advancing the cause of physical science.' He gave the Christmas Lectures on chemistry.

In June 1852, he sent a long paper to the ' Philosophical Magazine,' on the 'Physical Character of the Lines of Magnetic Force.' He begins with a note :—

'The following paper contains so much of a speculative and hypothetical nature that I have thought it more fitted for the pages of the "Philosophical Magazine" than for those of the "Philosophical Transactions." .... The paper, as is evident, follows series xxviii. and xxix., and depends much for its experimental support on the more strict results and conclusions contained in them.'

He made many reports to the Trinity House—among others:—on adulterated white-lead ; on oil in iron tanks ; on impure olive-oils ; on the Caskets lighthouse. The question of the use of Watson's electric light was first moved by a letter from Dr. Watson to the Trinity House.

1853, to the public, was a remarkable year in the life of Faraday. The chief work he did was for the public good. A popular error was opposed. Table-turning was rampant. In the judgment of some, the new motion showed a new force of nature. In the belief of others, it came from old Satanic action. Urged by many to say what he thought, he replied by making a few simple experiments. These showed how easily the uneducated judgment can be misled by the senses ; and how hard it is to shun errors in the interpretation of facts.

The wonders of human belief so astounded him, that he burst forth in a cry for the better education of the judgments around him. ' I declare that, taking the average of many minds that have recently come before me (and apart from that spirit which God has placed in each), and accepting for a moment that average as a standard, I should far prefer the obedience, affections, and instinct of a dog before it.'

The whole laboratory work of 1853 was done in six days in August and September.

In 1846 he had proved the influence and power of electricity and magnetism over a ray of light, and now again he hoped to reverse the order, and evolve or disturb these forces by the action of light. For this purpose he required a rock crystal which was in the British Museum, and he wrote—

TO THE TRUSTEES OF THE BRITISH MUSEUM.

'Royal Institution: August 20, 1853.

'My Lords and Gentlemen,—I am engaged in the investigation of a great object in natural science, namely, the relation of light to electricity and magnetism. I have advanced so far as to prove the influence and power of the two latter forces over a ray ("Philosophical Transactions," 1846, p. 1, &c.), and now hope to reverse the order and revolve or disturb these forces by the action of light. For these researches I need the use of a peculiar crystal of silica. I have sought for such, and have obtained some specimens, but they are too small to allow much hopes of success. In the British Museum I have found one, which for its clearness, size, position, plagiedral planes, and other circumstances, is eminently fitted for the research, and I cannot find such another. Under these circumstances, I take the liberty of making application for the loan of this crystal for the service of science. It is about thirteen and a half inches long, and four and a half inches in diameter, and is well known to Mr. Waterhouse and the attendants. I shall not in the slightest degree injure or even affect it, my only object being to pass a ray from the sun through it whilst it is surrounded by a helix, and in relation to a galvanometer. I need hardly say that I will take the utmost care of it; my willingness at all times to assist the Museum authorities in the *preservation* of the objects under their care, when they think that I can by my advice aid them in such matters, will, I hope, give assurance in that respect. I cannot tell for how long I may want it, for the experiments have to wait upon the sun. If I could transport the apparatus to the British Museum, I would propose that course, but the galvanometer is an especial instrument from Berlin, and requires fixing with the care of an astronomical instrument. I therefore hope that the trustees will permit me the use of this crystal in the Royal Institution. I would express my own deep thanks for such a favour, but that I feel it would be unmeet for me to offer private feelings or desires in such a case, and as I work for the pure good and advancement of science, I have no doubt that the trustees will do all that lies within their power to aid me. If by the use of the crystal an affirmative result were to be obtained, it would give the specimen a value far beyond any it could possess as a simple mineralogical illustration.

'I have the honour to be, with profound respect,

'My Lords and Gentlemen, your very humble and faithful servant,

'M. FARADAY.'

FARADAY TO SIR HENRY ELLIS.

'Royal Institution: August 20, 1853.

' My dear Sir Henry,—I beg, through you, as the proper channel, to make the enclosed application to the trustees of the British Museum. I have applied to Mr. Waterhouse for advice how to proceed, but I suppose I have been informal, for he does not encourage me. The trustees certainly ought to have the power, under sufficient precautions, to grant such a request as mine; for the British Museum is especially for the advancement of science. If they have not, I presume that some department of the Government has; but I think it can hardly be needful that I should make such application, or that I should move such bodies as the Royal Society, or the British Association, to make such application to a Secretary of State for a purpose so simple and, as it seems to me, so fit. Will you do me the favour to aid my object, and to let me know the result of my application. I am anxious, if possible, to make my experiments before the sun loses its power; otherwise, they will have to run on into next year.

' I am, my dear Sir Henry, your obliged and faithful servant,

' M. FARADAY.'

Sir H. Ellis answered:—

' The trustees, impressed by the importance of the object you have in view, have given instructions to deliver the crystal to you.'

FARADAY TO SIR HENRY ELLIS.

'Royal Institution : September 28, 1853.

'My dear Sir Henry,—I shall this day personally return the crystal to Mr. Waterhouse, and beg you will have the goodness to express my sincere thanks to the trustees of the British Museum for the favour granted me. I have optically examined the crystal ; and find it just what I wanted ; but from the delays which occurred, so much of the sunny weather has passed by that I have little hope now of any fit for my purpose this year. If, however, between this time and next summer, I am encouraged by results with other crystals, I may probably make application then for a second loan of the specimen.

'I am, my dear Sir Henry, your very obliged servant,
' M. FARADAY.'

The 'Athenæum' for July 2, 1853, contains a long letter from Faraday on table-moving. At a friend's house he made an experimental investigation into the results obtained by three skilful performers, and the beginning and ending of the account in the 'Athenæum' will sufficiently indicate the results he obtained.

'The following account of the methods pursued and the results obtained by Professor Faraday in the investigation of a subject which has taken such strange occupation of the public mind, both here and abroad, has been communicated to our columns by that high scientific authority. The subject was generally opened by Mr. Faraday in the " Times " of Thursday, June 30, it being therein intimated that the details were to be reserved for our this day's publication. The communication is of great importance in the present morbid condition of public thought,—when, as Professor Faraday says, the effect produced by table-turners has, without due inquiry, been referred to electricity, to magnetism, to attraction, to some unknown or hitherto unrecognised physical power able to affect inanimate bodies, to the revolution of the earth, and even to diabolical or supernatural agency :—and we are tempted to extract a passage from Mr. Faraday's letter to the " Times," which we think well worth adding to the experimental particulars and the commentaries with which he has favoured ourselves. " I have been," says the Professor, " greatly startled by the revelation which this purely physical subject has made of the condition of the public mind. No doubt there are many persons who have formed a right judgment or used a cautious reserve,—for I know several such, and public communications have shown it to be so ; but their number is almost as nothing to the great body who have believed and borne testimony, as I think, in the cause of error. I do not here refer to the distinction of those who agree with me and those who differ. By the great body, I mean such as reject all consideration of the equality of cause and effect,—who refer the results to electricity and magnetism, yet know nothing of the laws of these forces,—or to attraction, yet show no phenomena of pure attractive power,—or to the rotation of the earth, as if the earth revolved round the leg of a table,—or to some unrecognised physical force, without inquiring whether the known forces are not sufficient,—or who even refer them to diabolical or supernatural agency rather than suspend their judgment, or acknowledge to themselves that they are not learned enough in these 1853. matters to decide on the nature of the action. I think Æt.61–62. the system of education that could leave the mental condition of the public body in the state in which this subject has found it must have been greatly deficient in some very important principle.'

He ends thus: 'I must bring this long description to a close. I am a little ashamed of it, for I think, in the present age, and in this part of the world, it ought not to have been required. Nevertheless, I hope it may be useful. There are many whom I do not expect to convince; but I may be allowed to say that I cannot undertake to answer such objections as may be made. I state my own convictions as an experimental philosopher, and find it no more necessary to enter into controversy on this point than on any other in science, as the nature of matter, or inertia, or the magnetisation of light, on which I may differ from others. The world will decide sooner or later in all such cases, and I have no doubt very soon and correctly in the present instance. Those who may wish to see the particular construction of the test apparatus which I have employed, may have the opportunity at Mr. Newman's, 122 Regent Street. Further, I may say, I have sought earnestly for cases of lifting by attraction, and indications of attraction in any form, but have gained no traces of such effects.'

A few weeks later he writes a remarkable letter to his friend Professor Schönbein, in which his inner thoughts on this subject are seen.

FARADAY TO PROFESSOR SCHÖNBEIN.

'Royal Institution : July 25, 1853.

' My dear Schönbein,—I believe it is a good while since I had your last letter, but consider my age and weariness, and the rapid manner in which I am becoming more and more inert, and forgive me. Even when I set about writing, I am restrained by the consciousness that I have nothing worth communication. To be sure, many letters are written having the same character, but then there is something in the manner which makes up the value, and which, when I receive a letter from a kind friend such as you, often raises it in my estimation, far above what a mere reader would estimate it at. So you are going down the Danube, one point of which I once saw, and are about enjoying a holiday in the presence of fine nature. May it be a happy and a health-giving one, and may you return to your home loving it the better for the absence, and finding there all the happiness which a man sound both in mind and body has a right to expect on the earth.

' I have not been at work except in turning the tables upon the table-turners, nor should I have done that, but that so many inquiries poured in upon me, that I thought it better to stop the inpouring flood by letting all know at once what my views and thoughts were. What a weak, credulous, incredulous, unbelieving, superstitious, bold, frightened, what a ridiculous world ours is, as far as concerns the mind of man. How full of inconsistencies, contradictions, and absurdities it is. I declare that, taking the average of many minds that have recently come before me (and apart

from that spirit which God has placed in each), and
accepting for a moment that average as a standard, I
should far prefer the obedience, affections, and instinct
of a dog before it. Do not whisper this, however, to
others. There is One above who worketh in all things,
and who governs even in the midst of that misrule to
which the tendencies and powers of men are so easily
perverted.

'The ozone question appears, indeed, to have been
considerably illuminated by the researches in Bunsen's
laboratory. But why do you think it wonderful that
oxygen should assume an allotropic condition? We
are only beginning to enter upon the understanding of
the philosophy of molecules, and I think, by what you
say in former letters, that you are feeling it to be so.
Oxygen is to me of all bodies the most wonderful, as it
is to you. And truly the views and expectations of
the philosopher in relation to it would be as wild as
those of any table-turner, &c., were it not that the
philosopher has respect to the *laws* under which the
wonderful things that he acknowledges come to pass,
and to the never-failing recurrence of the *effect* when
the *cause* of it is present.

.   .   .   .

'My dear Schönbein, I really do not know what I
have been writing about, and I doubt whether I shall
re-read this scrawl, lest I should be tempted to destroy
it altogether. So it shall go as a letter, carrying with
it our kindest remembrances, and the sincerest affection
and esteem of yours ever truly,

'M. FARADAY.'

For the Institution in 1853 he gave a course of six lectures on static electricity.

Early in the year he gave a Friday discourse on observations on the magnetic force, and he gave the last lecture of the season on MM. Boussingault, Fremy, and Becquerel's experiments on oxygen.

At the Juvenile Lectures, which he gave at Christmas time on voltaic electricity, he spoke to the following effect, in consequence of a report that he had recanted the opinions on table-turning published in the ' Times.'

' In conclusion, I must address a few words to the intending philosophers who form the juvenile part of my audience. Study science with earnestness—search into nature—elicit the truth—reason on it, and reject all which will not stand the closest investigation. Keep your imagination within bounds, taking heed lest it run away with your judgment. Above all, let me warn you young ones of the danger of being led away by the superstitions which at this day of boasted progress are a disgrace to the age, and which afford astonishing proofs of the vast floods of ignorance overflowing and desolating the highest places.

' Educated man, misusing the glorious gift of reason which raises him above the brute, actually lowers himself below the creatures endowed only with instinct; inasmuch as he casts aside the natural sense, which might guide him, and in his credulous folly pretends to dissever and investigate phenomena which reason would not for a moment allow, and which, in fact, are utterly absurd.

' Let my young hearers mark and remember my words. I desire that they should dwell in their memory

as a protest uttered in this institution against the progress of error. Whatever be the encouragement it may
receive elsewhere, may we, at any rate, in this place,
raise a bulwark which shall protect the boundaries of
truth, and preserve them uninjured during the rapid
encroachments of gross ignorance under the mask of
scientific knowledge.'

In 1853, he gave five reports to the Trinity House—on
a comparison of a French lens and a Chance's lens; on
the lightning-rods at Eddystone and Bishop's lighthouses; on the ventilation of St. Catherine and the
Needles lighthouses, and that at Cromer; and on fog-
signals. A company was formed to apply Watson's
electric light to lighthouses, &c., but no trial of the light
took place.

II.

In 1853 he was made Foreign Associate of the Royal
Academy of Sciences, Turin, and Honorary Member of
the Royal Society of Arts and Sciences, Mauritius.

As director of the laboratory and superintendent
of the house, he received 300*l.* from the Royal Institution.

III.

A few characteristic letters written during these two
years remain.

Mr. Brande in 1852 gave his last lecture in the
theatre of the Royal Institution. He was told that the
secretary, Mr. Barlow, was instrumental in crushing the
expression of feeling of thankfulness for Mr. Brande's
long services, and he wrote to Mr. Faraday, who sent
the following reply :—

FARADAY TO PROFESSOR BRANDE.

'Royal Institution: May 6, 1852.

' My dear Brande,—Your informants have not done Barlow justice; perhaps they mistook him. I will endeavour to give you such an account as you ask me for, though doubting my memory. I was in the gallery at your last lecture; your audience were taken much by surprise by your farewell, and when you left the room, a member, I think Sir H. Hall, called on Mr. Barlow to take the chair, that the audience might give an expression of their feelings. Other persons spoke, and I think that several members thought they ought to have been informed by the managers of the coming resignation, and my impression was that they considered the managers ought to have done and said somewhat, and were hurt by the neglect. This, as *you* know, was impossible, because of the recent date of your announcement to the board, for the Monday following your lecture was the first monthly meeting after it. Mr. Barlow ventured to mention the recent circumstances, and the managers' intention to report on the following Monday your resignation, and their proposition to express their feelings by taking precisely the same steps as in the case of the resignation of Davy.

' Then expressions, very kind to you, were uttered, as was most natural after such a long term of what I may truly call affectionate relationship, accompanied by some vague propositions of a fellowship, a bust, or some other mark to be awarded as by a vote of those present. Mr. Barlow endeavoured to explain that that mixed meeting could not act or vote on a lecture day as a body of members, the act, charters, and by-laws being against it, but that they could give the expression of their conjoined opinion in any form they thought proper; whereupon a vote of thanks, moved and seconded by Sir Charles Clarke and Mr. John Pepys, was carried and communicated immediately by the former to you. The vote is also recorded in the printed notices, and I believe elsewhere, as in a report to the managers, but I am not sure about that. I cannot remember. On Monday, April 5, the managers made their report to the monthly meeting, and I had the honour of proposing you as the Honorary Professor of Chemistry. Several then spoke in the very highest terms of your long connection with the Royal Institution, and were glad to hear what the managers recommended. Several proposed some token of their feelings in which they could be joined personally, and Mr. John Pepys' generous mind was very forward in this; but a real obstruction was thrown in the way by one member proposing so many things that nothing was distinct; a chair, a scholarship, a bust presented to yourself, a portrait, a medal, were amongst them, and some members, including myself, had to remark upon the fitness of things. I recommended a committee, and it was understood, as I believe, that anything of the kind ought to be done, not as an act of the meeting of members acting for the *whole body* of members, but by a committee and subscription, as in other like cases; and I have been waiting to hear the formation of such a committee by those who seemed earnest for it.

' May 1st was the annual meeting; then also many kind expressions were uttered during the hour of waiting for the election of officers, your name being in the managers' list, but, as you know, nothing but the election and the visitors' report could then be taken as the business of the day. May 3 was the next monthly meeting, and then the election as Honorary Professor occurred. You ask me whether anything transpired. I cannot call to mind that any proposition (beyond what the managers had recommended) or any hint was made. I was still expecting the formation of a committee, but those who said most on the first occasion were not present.

'I have thus endeavoured to answer your inquiries, but feel I have not remembered the order of things clearly. Sir Charles Clarke was present on all the occasions, and he is one who could tell you what occurred, and whom I think you would feel you could trust. On the whole, I do not see how Mr. Barlow, when called upon, could act otherwise, and I know the impression on the minds of several who were present is *not* that which you have received. I shall say nothing to him or anybody else about your letter, but consider it at present quite confidential, as you desire; and I trust that you will soon hear enough from other parties to remove altogether and entirely the impressions you have received. It would be indeed a sad pity, if, after fifty years of kind and active association between the Royal Institution and yourself, the least uncomfortable feeling should remain as its result, and I cannot help saying that if I knew your informants I should feel very much inclined to speak to them as a justice due to Mr. Barlow and yourself conjointly.

'Ever, my dear Brande, yours faithfully,

'M. FARADAY.'

The substance of Mr. Brande's answer was : ' My   1852.
informants must have misunderstood or misinterpreted   Æt. 61.
Mr. Barlow's interference.'

FARADAY TO M. AUGUSTE DE LA RIVE.

Royal Institution : October 16, 1852.

' My dear de la Rive,—From day to day, and week
to week, I put off writing to you, just because I do not
feel spirit enough ; not that I am dull or low in mind,
but I am, as it were, becoming torpid, a very natural
consequence of that kind of mental fogginess which is
the inevitable consequence of a gradually failing
memory. I often wonder to think of the different
courses (naturally) of different individuals, and how
they are brought on their way to the end of this life.
Some with minds that grow brighter and brighter, but
their physical powers fail, as in our friend Arago, of
whom I have heard very lately, by a nephew who saw
him on the same day *in bed* and at *the Academy*, such
is his indomitable spirit. Others fail in mind first,
whilst the body remains strong. Others, again, fail in
both together, and others fail partially in some faculty
or portion of the mental powers, of the importance of
which they were hardly conscious until it failed them.
One may, in one's course through life, distinguish
numerous cases of these and other natures, and it is
very interesting to observe the influence of the respective circumstances upon the characters of the parties,
and in what way these circumstances bear upon their
happiness. It may seem very trite to say that content
appears to me to be the great compensation for these
various cases of natural change, and yet it is forced

upon me as a piece of knowledge that I have ever to call afresh to mind, both by my own spontaneous and unconsidered desires, and by what I see in others. No remaining gifts, though of the highest kind, no grateful remembrance of those which we have had, suffice to make us willingly content under the sense of the removal of the least of those which we have been conscious of. I wonder why I write all this to you? Believe me it is only because some expressions of yours at different times make me esteem you as a thoughtful man and a true friend. I often have to call such things to remembrance in the course of my own self-examinations, and I think they make me happier. Do not for a moment suppose that I am unhappy; I am occasionally dull in spirits, but not unhappy. There is a hope which is an abundantly sufficient remedy for that, and as that hope does not depend on ourselves I am bold enough to rejoice in that I may have it.

'I do not talk to you about philosophy, for I forget it all too fast to make it easy to talk about. When I have a thought worth sending you it is in the shape of a paper before it is worth speaking of; and after that it is astonishing how fast I forget it again. So that I have to read up again and again my own recent communications, and may well fear that as regards others I do not do them justice. However, I try to avoid such subjects as other philosophers are working at; and for that reason have nothing important in hand just now. I have been working hard, but nothing of value has come of it.

'Let me rejoice with you in the marriage of your daughter. I trust it *will be*, as I have no doubt it *has been*, a source of great happiness to you. Your son, too,

whenever I see him, makes me think of the joy he will
be to you. May you long be blessed in your children,
and in all the things which make a man truly happy
even in this life.

'Ever, my dear friend, yours affectionately,
'M. FARADAY.'

M. de la Rive's answer to this letter is essential for
the completion of the picture.

A. DE LA RIVE TO FARADAY.

'Genève, le 24 décembre 1852.

'Monsieur et très-cher ami,—Je n'ai pas répondu
plus tôt à votre bonne et amicale lettre, parce que
j'aurais voulu avoir quelque chose d'intéressant à vous
dire. Je suis peiné de ce que votre tête est fatiguée ;
céla vous est déjà arrivé quelquefois à la suite de vos
travaux si nombreux et si persévérants ; mais vous vous
rappelez qu'il suffit d'un peu de repos pour vous remettre
en très-bon état. Vous avez ce qui contribue le plus à
la sérénité de l'âme et au calme de l'esprit—une foi
pleine et entière et une conscience pure et tranquille,
qui remplit votre cœur des espérances magnifiques que
nous donne l'évangile. Vous avez en outre l'avantage
d'avoir toujours mené une vie douce et bien réglée,
exempte d'ambition, et par conséquent de toutes les
agitations et de tous les mécomptes qu'elle entraîne
après elle. La gloire est venue vous chercher malgré
vous ; vous avez su, sans la mépriser, la réduire à sa
juste valeur. Vous avez su vous concilier partout à la
fois la haute estime et l'affection de ceux qui vous con-
naissent.

'Enfin, vous n'avez été frappé jusqu'ici, grâce à la
bonté de Dieu, d'aucun de ces malheurs domestiques

qui brisent une vie. C'est donc sans crainte, comme sans amertume, que vous devez sentir approcher la vieillesse, en ayant le sentiment bien doux que les merveilles que vous avez su lire dans le livre de la nature doivent contribuer pour leur bonne part à en faire encore plus admirer et adorer le suprême Auteur.

'Voilà, très-cher ami, l'impression que votre belle vie m'a toujours fait éprouver ; et quand je la compare à nos vies agitées et si mal remplies, à tout cet ensemble de mécomptes et de douleurs dont la mienne en particulier a été abreuvée, je vous estime bien heureux, surtout parce que vous êtes digne de votre bonheur. Tout cela m'amène à penser au malheur de ceux qui n'ont pas cette foi religieuse que vous avez à un si haut degré.

.   .   .   .   .

'Je suis bien sûr que, malgré ce que vous me dites, vous saurez encore trouver quelque belle mine à exploiter dans ce riche domaine que vous cultivez avec tant d'ardeur et de succès. N'oubliez pas de m'en faire part à ce qui vient de vous, parce que c'est de vous avant tout, et ensuite parce que c'est toujours original et remarquable. Merci de toutes vos précédentes communications.

'Votre affectueux et bien dévoué

'A. De la Rive.'

FARADAY TO HIS NIECE MRS. DEACON.

'Royal Institution : July 23, 1853.

'My dear Caroline,—Yours to me arrived yesterday, and was very pleasant in the midst of the serious circumstances which had come over us, and of which you have no doubt heard by this time. I am always

cheered by your words, and it is well for us to hear a
remembrance of our hope rebounding from one to
another in these latter days, when the world is running
mad after the strangest imaginations that can enter the
human mind. I have been shocked at the flood of
impious and irrational matter which has rolled before
me, in one form or another, since I wrote my " Times "
letter ; and am more than ever glad that, as a natural
philosopher, I have borne my testimony to the cause of
common sense and sobriety. I have received letters
from the most learned and the highest, thanking me
for what I did.

' I cannot help thinking that the delusions of mind,
and the credulity which makes many think that supernatural works are wrought where all is either fancy or
knavery, are related to that which is foretold of the
latter days, and the prevalence of unclean spirits, which
unclean spirits are working in the hearts of man, and
not, as they credulously suppose, in natural things.
There is a good hope, however, which has no relation
to these things, except by its perfect separation from
them in all points, and which will not fail those who
are kept in it.

' Poor Mary ! But why poor ? She is gone in her
hope to the rest she was looking for, and we may
rejoice in her example as a case of the power of God,
who keeps those who look to Him in simplicity through
the faith that is in Christ. But her poor husband, and
her many children are deeply to be felt for, and you
also, and her father. We join in deep sympathy with
you all. . . .

' Do you see how crabbed my hand-writing has

become? The muscles do not obey as they used to do, but trip up or fall short of their intended excursions, and so parts of letters are wanting, or whole letters left out. You must guess it, and I know you will have a good will for the purpose.

. . . .

'And now, dear Caroline, with kindest remembrances to your husband, I must conclude, *not forgetting the maiden of the house.*

'Ever your affectionate uncle,

'M. FARADAY.'

His thoughts on scientific controversy were very strongly written to Professor C. Matteucci, who complained that Faraday had allowed the translation of Professor Müller's paper on animal electricity to be inscribed to him who had taken the greatest interest in the experiments.

FARADAY TO PROFESSOR C. MATTEUCCI.

'Royal Institution : March 3, 1853.

'My dear Matteucci,—I was quite startled the other day by the receipt of your letter (I mean the MS. one to myself), for my imperfect memory made me quite unaware that there was anything in Dr. B. Jones's *translation* of Müller's account of du Bois Raymond's experiments which could make it any source of annoyance or irritation beyond the original.

'I knew from matters reported in the "Comptes Rendus" and otherwise, that you and du Bois Raymond were in some degree antagonistically placed ; a thing very much to be regretted, but which often happens amongst the highest men in every department of science, and more often when there are two or three only that really pursue the subject than when there

are many.   Still I may truly say that when du Bois
Raymond was here, he never spoke of you in hard
terms or objectionably to me ; probably he avoided the
subject, but he did not embitter it.   Dr. Bence Jones's
translation was not completed, I think, in print, until
after he was gone, but of that I am not quite sure.
Being entirely unacquainted with German, I do not
know what either du Bois Raymond or Dr. Müller
may have said controversionally, but I concluded you
had borne with the work of the latter with that
patience which most men of eminence have to practise.
For who has not to put up in his day with insinuations
and misrepresentations in the accounts of his proceedings given by others, bearing for the time the
present injustice which is often unintentional, and often
originates in hasty temper, and committing his fame
and character to the judgment of the men of his own
and future time.

'I see that that moves you which would move me
most, namely, the imputation of a want of good faith,
and I cordially sympathise with any one who is so
charged unjustly.   Such cases have seemed to me
almost the only ones for which it is worth while
entering into controversy.   I have felt myself not un-
frequently misunderstood, often misrepresented, sometimes passed by, as in the cases of specific inductive
capacity, magneto-electric currents, definite electrolytic
action, &c. &c. ; but it is only in the cases where moral
turpitude has been implied, that I have felt called upon
to enter on the subject in reply.   I can feel with you in
the regret which you express (pp. 14, 15) at having to
write such a letter, and employ time in such a manner ;
and looking again at the abstract, can see how p. 23

and some other parts have made you think it necessary to do so ; but the letter being written, it will at all events have the good effect of collating dates both before and after the year 1842. Ultimately, this collation of dates is everything ; for in all matter of scientific controversy, the dates form the data upon which that final umpire appealed to (i.e. the scientific world) will judge.

' I am sorry the dedication annoys you : I suppose the Italian and the English feeling must differ in that respect. I do not like dedications, but I look upon them as honorary memberships, and not to be refused without something like an insult to the other parties concerned. In the chief number of cases in which I have been concerned, I have not been *asked* beforehand, and in all cases would rather not. We were bound by our duty to the members and to science to let du Bois Raymond (or any other like person) make his experiments here, and to the accident of his making them here is due the dedication itself, as the book says.

' These polemics of the scientific world are very unfortunate things ; they form the great stain to which the beautiful edifice of scientific truth is subject. *Are they inevitable?* They surely cannot belong to science itself, but to something in our fallen natures. How earnestly I wish, in all such cases, that the two champions were friends. Yet I suppose I may not hope that you and du Bois Raymond may some day become so. Well, let me be your friend, at all events ; and with the kindest remembrances to Madame Matteucci and yourself, believe me to be. my dear Matteucci,

' Ever very truly yours,

' M. FARADAY.'

To one who troubled him for his photograph, and would not be denied, he wrote :—

FARADAY TO DR. M———.

'Royal Institution : March 12, 1853.

'Dear Sir,—My words are *simple* and *correct.* I know that there are plenty of portraits : I do not know that there is a single likeness. I have compared the portraits with my face in the glass, and I cannot see a likeness in any one of them. Therefore, if I wished, I I could not send you one. But as I never help to publish either portrait or likeness, I cannot in any manner accede to your request. I think we may now consider this matter as finished.

'Very truly yours,
'M. FARADAY.'

I.

In 1854, though much original work was done, yet it fills but little space in the laboratory note-book. A long series of experiments were carried on, at the works of the Electric Telegraph Company, on submerged insulated wires of great length. He published a paper in the 'Philosophical Magazine' on this subject. He gave a lecture at the Royal Institution on it, and he sent an account of his experiments to M. de la Rive in a letter which also shows his loving nature.

FARADAY TO MONSIEUR AUG. DE LA RIVE.

'Royal Institution : January 28, 1854.

'My dear and kind Friend,—It seems a very long time since I wrote to or heard from you, but I have no doubt it has been my own fault. I often verify to myself the truth of the old school copy, " Procrastination is the thief of time;" and when I purpose to write, it seems to me as if my thoughts now were hardly worth utterance to the men of persisting intellect and strength. But there are ties besides those of mere science and worldly relationship, and I venture to think I have some such with you. These I cannot easily relinquish, for they grow dearer as other more temporal things dissolve away; and though one cannot talk so often or so glibly about them, because of their far more serious character, still from time to time we may touch these chords, and I shall think it a happiness whilst they respond and vibrate between us. Such ties exist but in few directions, but they are worth all the rest.

'I had a word from Schönbein a little while ago, and he called you to mind by speaking of his daughter, who was, I think, then with you, and it called up afresh the thoughts of the place when, very many years ago, I first saw it and your father, 1814 or '15, but the remembrances of that time are very shadowy with me. There came up the picture of the time when I and my wife were there with you and your happy family, and a strong thought of the kindness I have had from your house through two generations, and now comes the contemplation of these generations passing away. Surely, though we have both had trials, and deep ones, yet we have also had great mercies and goodness shown us; above all, the *great hope*. May the year that we have entered be full of peace to you and sweet pleasure among your children.

'I have lately had a subject brought before me in electricity, full of interest. My account of it is in the printer's hands, and when I receive it I will send you a copy by post.   Briefly, it is this.   Copper telegraph wires are here covered perfectly with gutta-percha, so that hundreds of miles may be immersed in water, and yet a very small discharge through the gutta-percha occurs, when a very intense voltaic battery (300 or 400 pairs of plates) is connected with it.   100 miles of such wire in water with the two exposed extremities insulated, can be charged by one pole of a voltaic battery, and after separation from the battery for five or ten minutes, will give a shock or a current to the body or a galvanometer, or fire gunpowder, or effect other electric actions, either static or dynamic.   The 100 miles is, in fact, an immense Leyden jar, and because the copper is $\frac{1}{16}$th of an inch in diameter, and the gutta-percha $\frac{1}{10}$th of an inch thick or $\frac{1}{16}$th of an inch extreme diameter, the surface of the copper or inner lining of the jar is equal to 8,270 square feet, and the outer coating or water surface equal to 33,000 square feet.   But besides this fact of a charge given, kept, and then employed, such a wire in water has its power of conveying electricity wonderfully affected; not its final power, for that is the same for an equal length of the wire in air or in water, but its power in respect of brief currents or waves of electric force, even to the extent of making the time occupied in the transmission vary as 100 to 1 or more.   In a few days you shall have the account.   I do not know whether I have told you yet of the pleasure I have had in your vol. i.,[1] but I long for vol. ii.   Many thanks for all your kindness in it, and on every occasion.

'Ever, my dear de la Rive, yours,

'M. Faraday.'

---

[1] On Electricity.

This year is memorable at the Royal Institution for a lecture which Faraday gave on mental education, or as he also called it, on deficiency of judgment, and on the means of improving it. This formed one of an afternoon course of lectures on education, and Prince Albert came to it. He began the lecture with some very remarkable words; they are in the highest degree characteristic of his mind: 'Before entering upon this subject, I must make one distinction which, however it may appear to others, is to me of the utmost importance. High as man is placed above the creatures around him, there is a higher and far more exalted position within his view; and the ways are infinite in which he occupies his thoughts about the fears, or hopes, or expectations of a future life. I believe that the truth of that future cannot be brought to his knowledge by any exertion of his mental powers, however exalted they may be; that it is made known to him by other teaching than his own, and is received through simple belief of the testimony given. Let no one suppose for a moment that the self-education I am about to commend, in respect of the things of this life, extends to any considerations of the hope set before us, as if man by reasoning could find out God. It would be improper here to enter upon this subject further than to claim an absolute distinction between religious and ordinary belief. I shall be reproached with the weakness of refusing to apply those mental operations which I think good in respect of high things to the very highest. I am content to bear the reproach. Yet even in earthly matters I believe that "the invisible things of Him from the creation of the world are clearly seen, being understood by the things that are

made, even His eternal power and Godhead," and I 1854.
have never seen anything incompatible between those Æt.62–63.
things of man which can be known by the spirit of
man which is within him, and those higher things
concerning his future, which he cannot know by that
spirit.' And then he proceeds to lecture on *deficiency
of judgment.*

He ends his lecture by saying : ' My thoughts would
flow back amongst the events and reflections of my past
life, until I found nothing present itself but an open
declaration—almost a confession—as a means of performing the duty due to the subject and to you.'

When reprinting these thoughts on mental education, he said : ' They are so immediately connected in
their nature and origin with my own experimental life,
considered either as cause or consequence, that I have
thought the close of this volume (of " Researches on
Chemistry and Physics ") not an unfit place for their
reproduction.'

In addition to the Friday evening discourse on electric induction, he gave the last Friday discourse on
magnetic hypothesis. He ended it thus : ' Our varying
hypotheses are simply the confessions of our ignorance
in a hidden form ; and so it ought to be, only the
ignorance should be openly acknowledged.'

He gave the Christmas Lectures on combustion.

At the end of this year he sent a long paper to the
' Philosophical Magazine ' on some points of magnetic
philosophy. He begins saying :—' Within the last
three years I have been bold enough, though only as
an experimentalist, to put forth new views of magnetic
action in papers having for titles " On Lines of Magnetic
Force," " Phil. Trans.," 1852 ; and " On Physical Lines

of Magnetic Force," "Phil. Mag.," 1852. I propose to call the attention of experimenters in a somewhat desultory manner to the subject again, both as respects the deficiency of the present physical views and the possible existence of lines of physical force.'

He ends his paper saying:—' I have no clear idea of the physical condition constituting the charged magnetic state, i.e. the state of the source of magnetic power, or of the coercitivity by which that state is either resisted in its attainment or sustained in its permanent condition, for the hypotheses as yet put forth give no satisfaction to my mind. I profess rather to point out the difficulties in the way of the views which are at present somewhat too easily accepted, and to shake men's minds from their habitual trust in them; for next to developing and expounding, that appears to me the most useful and effectual way of really advancing the subject: it is better to be aware or even to suspect we are wrong, than to be unconsciously or easily led to accept an error as right.'

The two next letters to Professor de la Rive are on the production of induction currents in liquids, on which Faraday made some experiments this year.

FARADAY TO PROFESSOR DE LA RIVE.

'Royal Institution: March 1, 1854.

' My dear Friend,—Your kindness and invitation move our hearts to great thankfulness youwards: but they cannot roll back the years, and give us the strength and ability of former times. We are both changed— my wife even more than I, for she is indeed very infirm in her limbs; nor have I much expectation that in that respect she will importantly improve; but we are both very thankful for each other's company, and for the abundant blessing God has granted to us. I do not think it probable that either of us shall cross the sea this year, or move a hundred miles from home, but we shall often during the summer recall to mind your very pleasant invitation.

'Your volume and the new matter I shall look forward to with eagerness. My little report I have no doubt you have received ere this. You will there perceive how much the induction you referred to in your letter has to do with the phenomena described.

'Now in reference to your questions. And first, whether I have ever obtained induction currents through liquids not being metals? I have not worked on the subject since 1832. At that time I obtained *no current* with a tube of sulphuric acid (Experimental Research, 200), but the current obtained in metals passed through liquids (Experimental Research, 20). I should not at all despair of obtaining the current by the use of electro-magnets and thick wire galvanometers (3,178), but I never have obtained them.

'With regard to the second question, I have never seen any reason to withdraw from the opinion I formed in the year 1834, that water and such liquids could conduct a very feeble portion of electricity without suffering decomposition. I venture to refer you to the paragraphs in the "Experimental Researches," namely, 968 to 973, also 1017 and 1032. I have never contested the point, because, having once advanced it, I have not since found any reason to add or alter, and I left it to make its way. You will find at the end of paragraph 984 reference to a point which has always had great weight with me. When electrolytes are solid, as in the case of nitre or chloride of sodium at common temperatures, or water at or below 0° F. ; and when, according to all appearance, they *cannot* conduct as electrolytes, they still can conduct electricity of high tension, as is shown at par. 419 to 430. If they have this power to such a considerable degree, with electricity able to open the gold leaves, it is almost certain that they have it to a certain degree with electricity of lower tension ; and if the solid electrolytes have such power, I cannot see any reason why their liquefaction should take it away. It would seem rather unphilosophical to admit it for the solid, and then without proof to assume that it is absent in the liquid : for my part, I think the proof is all the contrary way. The power seems to be present in a very low degree, but I think it is there. So much for that matter.

' If I were in your company I should have a long chat with you about Pelago's experiments. I cannot understand them as to any new principle that is involved in them ; and if there be not a new principle I fear they are only mistakes, i.e. imperfect forms of old results where the two developed forces are beforehand present. I cannot conceive it possible that if a sphere (metallic) of three inches diameter, be inside a metallic sphere of twelve feet (or any other) diameter, and touching its side, its mere removal into the centre of the larger sphere, or any other position in it, will cause any electricity to appear.

' Adieu, my dear friend, for the present.

'Ever affectionately yours,

'M. FARADAY.'

FARADAY TO PROFESSOR DE LA RIVE.

'Royal Institution : March 7, 1854.

' My dear Friend,—Your question whether I have
ever succeeded in producing induction currents in other
liquids than mercury or melted metals, as, for instance,
in acid or saline solutions, has led me to make a few
experiments on the subject, for though I believe in the
possibility of such currents, I had never obtained affirmative results. I have now procured them, and send
you a description of the method pursued. A powerful
electro-magnet of the horse-shoe form was associated
with a Grove's battery of twenty pairs of plates. The
poles of the magnet were upwards, their flat end faces
being in the same horizontal plane. They are 3·5
inches square, and about 6 inches apart. A cylindrical
bar of soft iron, 8 inches long and 1·7 in diameter, was
employed as a keeper or submagnet. The cylindrical
form was adopted, first, because it best allowed of the
formation of a fluid helix around it, and next because
when placed on the poles of the magnet, and the battery
connections made and broken, the magnet and also the
keeper rises and falls through much larger variations
of power, and far more rapidly than when a square or
flat-faced keeper is employed ; for the latter, if massive,
has, as you know, its power of sustaining the magnetic
conditions of the magnet in a very great degree when
the battery connection is broken. A fluid helix was
formed round this keeper, having twelve convolutions,
and a total length of 7 feet ; the fluid was only 0·25 of
an inch in diameter, the object being to obtain a certain
amount of intensity in the current by making the

inductive excitement extend to all parts of that great
length rather than to produce a quantity current by
largeness of diameter, i.e. by a shorter mass of fluid.

'This helix was easily constructed by the use of 8·5
feet of vulcanised caoutchouc tube having an internal
diameter of 0·25, and an external diameter of 0·5 of an
inch. Such a tube is sufficiently strong not to collapse
when placed round the iron cylinder. The twelve convolutions occupied the interval of 6 inches, and two
lengths of 9 inches each constituted the ends. This
helix was easily and perfectly filled by holding it with
its axis perpendicular, dipping the lower end into the
fluid to be used, and withdrawing the air at the upper ;
then two long clean copper wires 0·25 of an inch in
diameter were introduced at the ends, and being thrust
forward until they reached the helix, were made secure
by ligaments, and thus formed conductors between the
fluid helix and the galvanometer. The whole was
attached to a wooden frame, so as to protect the helix
from pressure or derangement when moved to and fro.
The quantity of fluid contained in the helix was about
3 cubic inches in the length of 7 feet. The galvanometer was of wire 0·033 of an inch in diameter, and 164
feet in length, occupying 310 convolutions. It was
18 feet from the magnet, and connected with the helix
by thick wires dipping into cups of mercury. It was
in the same horizontal plane with the magnetic poles,
and very little affected by direct action from the latter.

'A solution formed by mixing one volume of strong
sulphuric acid, and three of water was introduced into
the helix tube. The iron keeper placed in the helix,
and the whole adjusted on the magnetic poles in such
a position that the ends of the copper connectors in

the tube were above the iron cylinder or keeper, and were advanced so far over it as to reach the perpendicular plane passing through its axis. In this position the lines of magnetic force had no tendency to excite an induced current through the metallic parts of the communication. The outer ends of the copper terminals were well connected together, and the whole left for a time, so that any voltaic tendency due to the contact of the acid and copper might be diminished or exhausted.

' After that the copper ends were separated, and the connections with the galvanometer so adjusted that they could be in an instant either interrupted or completed, or crossed at the mercury cups. Being interrupted, the magnet was excited by the full force of the battery, and thus the direct magnetic effect on the galvanometer was observed. The helix had been so arranged that any current induced in it should give a deflection in the contrary direction to that thus caused directly by the magnet, that the two effects might be better separated. The battery was then disconnected, and when the reverse action was over the galvanometer connections were completed with the helix. This caused a deflection of only 2°, due to a voltaic current generated by the action of the acid in the helix on the copper ends. It showed that the connection throughout was good, and being constant in power caused a steady deflection, and was thus easily distinguished from the final result. Lastly, the battery was thrown into action upon the magnet, and immediately the galvanometer was deflected in one direction, and upon breaking battery contact it was deflected in the other direction ; so that by a few alterna-

tions considerable swing could be imparted to the
needles. They moved also in that particular manner
often observed with induced currents, as if urged by an
impact or push at the moments when the magnet was
excited or lowered in force ; and the motion was in the
reverse direction to that produced by the mere direct
action of the magnet. The effects were constant. When
the communicating wires were crossed they again occurred, giving reverse actions at the galvanometer.
Further proof that they were due to currents induced
in the fluid helix was obtained by arranging one turn
of a copper wire round the iron core or keeper, in the
same direction as that of the fluid helix, and using one
pair of plates to excite the magnet ; the induced current
caused in the copper wire was much stronger than that
obtained in the fluid, but it was always in the same
direction.

'After these experiments with the highly conducting
solution, the helix was removed, the dilute acid poured
out, a stream of water sent through the helix for some
time, distilled water then introduced, and allowed to
remain in it awhile, which being replaced by fresh
distilled water, all things were restored to their places
as before, and thus a helix of pure water was submitted
to experiment. The direct action of the magnet was
the same as in the first instance, but there was no appearance of a voltaic current when the galvanometer
communications were completed. Nor were there any
signs of an induced current upon throwing the magnet
into or out of action. Pure water is too bad a conductor to give any sensible effects with a galvanometer
and magnet of this sensibility and power.

'I then dismissed the helix, but placing the keeper

on the magnetic poles, arranged a glass disc under it, and filled the dish with the same acid solution as before. So that the liquid formed a horizontal fluid disc, six inches in diameter nearly, an inch deep, and within 0·25 of an inch of the keeper. Two long clean platinum plates dipped into this acid on each side of the keeper, and parallel to it, and were at least five inches apart from each other; these were first connected together for a time, that any voltaic tendency might subside, and then arranged so as to be united with the galvanometer when requisite, as before. Here the induced currents were obtained as in the first instance, but not with the same degree of strength. Their direction was compared with that of the current induced in a single copper wire passed between the fluid and the keeper, the magnet being then excited by one cell, and was found to be the same. However, here the possibility exists of the current being in part or altogether excited upon the portions of the wire conductors connected with the platinum plates; for as their ends bend to go beneath the keeper, and so into the circuit of magnetic power formed by it and the magnet, they are subject to the lines of force in such a position as to have the induced current formed in them; and the induced current can obtain power enough to go through liquid, as I showed in 1831. But as the helix experiment is free from this objection, I do not doubt that a weak induced current occurred in the fluid in the dish also.

'So I consider the excitement of induction currents in liquids not metallic as proved; and as far as I can judge they are proportionate in strength to the conducting powers of the body in which they are generated. In the dilute sulphuric acid they were of course stronger than they appeared by the deflection to be ; because they had first to overcome the contrary deflection which the direct action of the magnet was able to produce. The sum of the two deflections, in fact, expressed the force of the induced current. Whether the conduction, by virtue of which they occur, is electrolytic in character or conduction proper, I cannot say. The present phenomena do not aid to settle that question, because the induced current may exist by either one or the other process. I believe that conduction proper exists, and that a very weak induction current may pass altogether by it, exerting for the time only a tendency to electrolysis : whilst a stronger current may pass partly by it, and partly by full electrolytic action.

'I am, my dear friend, ever most truly yours,

'M. FARADAY.'

FARADAY TO DE LA RIVE.

'Royal Institution : March 8, 1854.

'My dear de la Rive,—I send you the (above) enclosed letter in such shape that you may publish it if you think it worth while. It has been copied so as to be a little better in writing than if you had had the original. I wish I could have written it in French. As the experiments arose out of your question, I send the matter to you first. If you publish it in the " Bibliotheque," then I shall afterwards give my rough copy to the " Philosophical Magazine " as the translation from your journal.

'If you should not find it expedient to print it, then I would alter the heading a little, and send it to the

" Philosophical Magazine" as original. Do exactly as 1854.
you like with it.   Æт. 62.

'Ever, my dear friend, yours affectionately,

'M. FARADAY.'

He sent five reports to the Trinity House, one of
which, in two parts, was on Dr. Watson's electric light
(voltaic), and on Professor Holmes's electric light
(magneto-electric). The conclusion was that he could
not recommend the electric light, and that it had better
be tried for other than lighthouse uses first. To Dr.
Watson he wrote that he ' could not put up in a lighthouse what has not been perfectly established beforehand, and is only experimental.'

II.

He was made Corresponding Associate of the Royal
Academy of Sciences, Naples.

III.

Several interesting letters this year show Faraday's
character. His kindness and nobleness come out everywhere. The most characteristic letter is one written in
answer to a question from the Parliamentary Committee
of the British Association, Whether any and what measures could be adopted by the Government or the
Legislature, to improve the position of science, or of the
cultivators of science, in this country? This was dated
March 8, 1854, and signed ' Wrottesley, Chairman.'

FARADAY TO LORD WROTTESLEY.

'Royal Institution : March 10, 1854.

' My Lord,—I feel unfit to give a deliberate opinion
on the course it might be advisable for the Government
to pursue if it were anxious to improve the position of

science and its cultivators in our country. My course of life, and the circumstances which make it a happy one for me, are not those of persons who conform to the usages and habits of society. Through the kindness of all, from my sovereign downwards, I have that which supplies all my need ; and in respect of honours, I have, as a scientific man, received from foreign countries and sovereigns, those which, belonging to very limited and select classes, surpass in my opinion anything that it is in the power of my own to bestow.

'I cannot say that I have not valued such distinctions ; on the contrary, I esteem them very highly, but I do not think I have ever worked for or sought after them. Even were such to be now created here, the time is past when these would possess any attraction for me ; and you will see therefore how unfit I am, upon the strength of any personal motive or feeling, to judge of what might be influential upon the minds of others. Nevertheless, I will make one or two remarks which have often occurred to my mind.

' Without thinking of the effect it might have upon distinguished men of science, or upon the minds of those who, stimulated to exertion, might become distinguished, I do think that a Government should *for its own sake*, honour the men who do honour and service to the country. I refer now to honours only, not to beneficial rewards ; of such honours I think there are none. Knighthoods and baronetcies are sometimes conferred with such intentions, but I think them utterly unfit for that purpose. Instead of conferring distinction, they confound the man who is one of twenty, or perhaps fifty, with hundreds of others. They depress rather than exalt him, for they tend to lower the especial

distinction of mind to the commonplaces of society. An
intelligent country ought to recognise the scientific men
among its people as a class.   If honours are conferred
upon eminence in any class, as that of the law or the
army, they should be in this also.   The aristocracy of
the class should have other distinctions than those of
lowly and high-born, rich and poor, yet they should be
such as to be worthy of those whom the sovereign and
the country should delight to honour, and being
rendered very desirable and even enviable in the eyes
of the aristocracy by birth, should be unattainable
except to that of science.   Thus much I think the
Government and the country ought to do for their
own sake and the good of science, more than for the
sake of the men who might be thought worthy of such
distinction.   The latter have attained to their fit place,
whether the community at large recognise it or not.

' But besides that, and as a matter of reward and
encouragement to those who have not yet risen to great
distinction, I think the Government should in the very
many cases which come before it, having a relation to
scientific knowledge, employ men who pursue science
provided they are also men of business.   This is perhaps
now done, to some extent, but to nothing like the
degree which is practicable with advantage to all
parties.   The right means cannot have occurred to a
Government which has not yet learned to approach and
distinguish the class as a whole.

' At the same time, I am free to confess that I am
unable to advise how that which I think should be may
come to pass.   I believe I have written the expression
of feelings rather than the conclusions of judgment, and
I would wish your Lordship to consider this letter as

private rather than as one addressed to the Chairman of a Committee.

'I have the honour to be, my Lord, your very faithful servant,　　　　　　　　'M. FARADAY.'

FARADAY TO PROFESSOR SCHÖNBEIN.

'Royal Institution: May 15, 1854.

'My dear Schönbein,—Your letters stimulate me by their energy and kindness to write, but they also make me aware of my inability, for I never read yours even for that purpose without feeling barren of matter and possessed of nothing enabling me to answer you in kind. And then, on the other hand, I cannot take yours, and think it over, and so generate a fund of philosophy as you do, for I am now far too slow a man for that. What is obtained tardily by a mind not so apt as it may have been is soon dropped again by a failing power of retention. And so you must just accept the manifestation of old affection and feeling in any shape that it may take, however imperfect.

'I made the experiments on the dahlia colour, which you sent me, and they are very beautiful. Since then, I have also made the experiment with ink and carbonic acid (liquid), and succeeded there also to the extent you described. I had no reason to expect, from what you said, that dry ink would lose its colour, but I tried the experiment, and could not find that the carbonic acid bath had power to do that. Many years ago I was engaged on the wonderful power that water had when it became ice of excluding other matters. I could even break up compounds by cold. Thus if you prepare a thin glass test tube about the size of the thumb, and a feather so much larger that when in the tube and twirled about it shall rapidly brush the sides, if you take some dilute sulphuric acid so weak that it will easily freeze at 0° Fahrenheit, and putting that into the tube with the feather; if, finally, you put all into a good freezing mixture of salt and snow, and whilst the freezing goes on, you rotate the feather continually and quickly, so as to continually brush the interior surface of the ice formed, clearing off all bubbles, and washing that surface with the central liquid, you may go on until a half or two-thirds or more of the liquid is frozen, and then pouring out the central liquid you will find it a concentrated solution of the acid. After that, if you wash out the interior of the frozen mass with two or three distilled waters, so as to remove all adhering acid, and then warm the tube by the hand, so as to bring out the piece of ice, it upon melting will give you pure water, not a trace of sulphuric acid remaining in it. The same was the case with common-salt solution, sulphate of soda, in alcohol, &c., and, if I remember rightly, even with some solid compounds of water. I think I recollect the breaking up of crystals of sulphate of soda by cold, and I should like very much now to try the effect of a carbonic acid bath on crystals of sulphate of copper. So it strikes me that in the effect of the cold on the colourless dahlia solution, the reappearance of the colour may depend upon the separation of the sulphurous acid from the solidifying water.

. . . . . .

' I think some of my letters must have missed—you scold me so hard. As I cannot remember what I have sent or said, I am obliged to enter in a remembrancer the letters written or received, and, looking to it, I find the account thus   .   .   .   .

1854.
_Æt._ 62.

.   .   .   .   .   .

and considering that I have little or nothing to say, and you are a young man in full vigour, that is not so very bad an account ; so be gentle with your failing friend.

'Ever, my dear friend, affectionately yours,

'M. FARADAY.'

On a question relating to the Queen's yacht, he thus writes to the Hon. Captain Jos. Denman.

FARADAY TO THE HON. CAPTAIN JOS. DENMAN.

'Royal Institution : May 27, 1854.

' My dear Sir,—Your letter is full of interest, and I feel great delight that any conversation in which I had part should be connected with so just an application of the principles of natural philosophy, as has been made by His Royal Highness Prince Albert, in the cases of the paddle-wheel and the propeller.

' You will be aware, from the communication of his Royal Highness, that all practical result may be referred to the following facts. A disc when rotating resists any force tending to alter its place, so as to change the plane of its rotation, far more than if the disc were not rotating, and the resistance is the greater as the body is heavier, as the parts have greater velocity or momentum, and therefore as they are further from the axis of rotation, and as the change of place is greater. Now the force of the paddle-wheels, and their positions in relation to a steam ship, are such that they cannot but affect its rolling, and their tendency will be to diminish it. You will understand that the endeavour is not to preserve any particular plane as regards the horizon, but *that* in which the disturbing force finds

the rotating disc ; so if a wave causes the vessel to roll,
the revolving bodies will tend to resist this roll; as
the vessel endeavours to recover itself the tendency will
be to resist the recovery also ; but, on the whole, the
rolling will be obstructed and diminished.   I have
always considered that paddle-wheels resist and diminish
rolling by the hold the descending side takes (like a
hand in swimming) upon the water ; but I have not the
slightest doubt now, that they will act by the effect His
Royal Highness has pointed out.   What the proportion
may be I cannot say, or to what extent the weight of
seventy tons disposed in forms about thirty-two feet in
diameter, and revolving once in two seconds, would
affect a ship of 2300 tons.   But I should expect it
would be very appreciable, and should not be surprised
if it may form a considerable part of any superiority
which paddle-wheels have over screws.

' The screw you refer to, though it would revolve
with twice the velocity of the paddle-wheels, has only
half their diameter and a third of their weight ; so
that it would present much less resistance to change of
plane than the latter.   Besides this, it is at the extremity
of the vessel, and therefore perhaps six or eight times
as far from the horizontal transverse line about which
the ship tends to revolve when pitching as the paddle-
wheels are from the horizontal longitudinal line about
which the ship tends to revolve when rolling ; for the
short motions of the roll will be much more resisted
than the long motion of the pitch, because the place of
rotation, in the first case, is more quickly changed.   I
do not think that the screw would tend to increase
rolling otherwise than as it would replace the paddle-
wheels, which tend to diminish it.

'The suggestion of His Royal Highness in regard to a central fly-wheel is highly philosophic, and perfectly justified by natural principles. At the same time, I cannot undertake to say what amount of effect it would produce in any given case. Still the experiment could be made so simply and progressively that I think any marine engineer could ascertain the point practically in a very few days.

'Suppose a boat with a heavy disc or fly-wheel fitted up in the middle, this being attached by running bands to an axle and handles in the fore or aft parts, so that a man (or two men if needful) could get the fly into rapid rotation, the boat being of such size that a third person, standing across or from side to side, could by the action of his limbs sway her right and left: he might do this when the fly is still, and also when in quick motion : he would soon find the resistance to his efforts in the latter case, and then a judgment might be formed as to the result of a *larger experiment* and as to the application to a ship. If more convenient, two fly-wheels might be used, one on each side of the boat, and the gear and men be in the middle ; but the first experiment ought to be made with a boat that can be easily and quickly rocked, or the results will not be so instructive as they might be.

'Though I have spoken thus far of a disc revolving in a vertical plane, yet it is of course evident that a horizontal or any other plane may be selected, provided that the axis of rotation is perpendicular to the length of the boat.

'Supposing that a great disc or fly-wheel were revolving in the inside of a vessel parallel to and in the same direction as the paddle-wheels, and a wave were to affect the vessel, rolling her, so as to depress the starboard side, the resistance set up by the disc would not be direct, but would have an oblique result, tending to turn the ship's head to starboard. Has anything of this kind been distinguished by the man at the wheel? Probably he could not tell it from the effect due to immersion of the starboard paddle-wheel. In the boat experiment it ought to be sensible.

'I am, my dear Sir, very truly yours,

'M. FARADAY.'

A letter to Professor de la Rive shows how he felt his power failing.

FARADAY TO PROFESSOR DE LA RIVE.

'Royal Institution: May 29, 1854.

'My dear Friend,—Though feeling weary and tired, I cannot resist any longer conveying to you my sincere thanks (however feebly) for the gift of your work in French. I have delayed doing so for some time, hoping to be in better spirits, but will delay no longer. For delighted as I have been in the reading of it, my treacherous memory begins to let loose that which I gained from it; for when I read some of the summaries a second time, I am surprised to find them there, and then slowly find that I had read them before. The power with which you hold the numerous parts of our great department of science in your mind is to me most astonishing and delightful, and the accounts you give of the researches of the workers, and especially those of Germany, are exceedingly valuable and interesting to me. May you long enjoy and use this great power for the good of us all. We shall long for

z 2

the second volume ; but we must have patience, for it is a great work that you are engaged in.

' You sent me also the numbers of the "Bibliotheque" for January, February, and March, and then again your kindness to me is deeply manifested, and with me is deeply felt ; but do not trouble yourself to send me the succeeding numbers, for I have the work here, and see it with great interest, for it is to me a channel for much matter that otherwise would escape me altogether. I wish I could send you matter oftener, but my wishes far antimeasure my abilities. My portfolio contains many plans for work, but I get tired with ordinary occupation, and then my hands lie idle.

' Your theoretical views, from p. 557 and onwards, have interested me very deeply, and I am glad to place them in my mind, by the side of those ideas which seem to aid discovery and development by suggesting analogies and crucial experiments, and other forms of test for the views which arise in the mind as vague shadows, however they may develope into brightness. I have always a great difficulty about hypotheses, from the necessity one is under of holding them loosely, and suspending the mental decision. I do not know whether I am right in concluding that your hypothesis supposes that there can only be a few atoms in each molecule, and that these are arranged as a disc, or, at all events, disc-fashioned, i.e. in the same plane. It seems to me that if we consider a molecule in its three dimensions, it will be necessary to consider the atoms as all having their axis in planes parallel to one only of these directions, however numerous these atoms may be. I speak, of course, of those bodies which you consider as naturally magnetic, page 571. Perhaps when

I get my head a little clearer, I may be able to see more clearly the probable arrangements of many atoms in one molecule. But for the present I must refrain from thinking about it.

'Our united, kindest remembrance. Ever, my dear friend, your faithful,   'M. FARADAY.'

FARADAY TO PROFESSOR SCHÖNBEIN.

'Royal Institution : September 15, 1854.

'My dear Schönbein,—Just a few scattered words of kindness, not philosophy, for I have just been trying to think a little philosophy (magnetical) for a week or two, and it has made my head ache, turned me sleepy in the day-time as well as at nights, and, instead of being a pleasure, has for the present nauseated me. Now you know that is not natural to me, for I believe nobody has found greater enjoyment in physical science than myself; but it is just weariness, which soon comes on, but I hope will soon go off by a little rest.

'The July letter was a great delight ; both your kindness and your philosophy most acceptable and refreshing. I hope to get your paper translated, but there is a great deal of vis inertiæ in our way, and I cannot overcome it, as I would wish to do. It is the more difficult for me to criticise it, because I feel a good deal of it myself, and am known to withdraw from the labour and responsibilities of scientific work, and this makes me very glad that you have got hold of Liebig, for I hope he will act in developing your ozone views.

'You give a happy account of your family. You

are a happy man to have such a family, and you are
happy in the temperament which fits you for the enjoyment of it. May God bless every member of it and
yourself with a cheerful and relying spirit and love to
each other. Remember us to them all.

'Ever, my dear friend, affectionately yours,

'M. FARADAY.'

FARADAY TO DR. TYNDALL.[1]

'Royal Institution : November 11, 1854.

'Many thanks, my dear Tyndall, for your kind letter,
which I have just received. I was anxious about you,
thinking you might be confined at home by a little
indisposition (as you would call it) and writing, and
should probably have called to-day in the evening.
Now I shall rest, knowing how it is, and I hope you
will enjoy the weather, and the quietness, and the time
of work, and the time of play, finding them all
ministrants to your health and contented happiness.

'Here we jog on, and I have just undertaken the
Juvenile Lectures at Christmas, thinking them the easiest
thing for me to do. Reading Matteucci carefully, and
also an abstracted translation of Van Rees' paper, is my
weighty work, and because of the call it makes on
memory I have now and then to lay them down and
cease till the morrow. I think they encourage me to
write another paper on lines of force, polarity, &c., for
I was hardly prepared to find such strong support in
the papers of Van Rees and Thomson for the lines as
correct representants of the power and its direction;
and many old arguments are renewed in my mind by

---

[1] In 1853, Dr. Tyndall became Professor of Physics in the Royal
Institution.

these papers. But we shall see how the maggot bites   1855.
presently; and as I fancy I have gained so much by   Æt.63–64.
waiting, I may perhaps wait a little longer.

'Ever, my dear Tyndall, yours truly,

'M. FARADAY.'

The year 1855 brought the series of experimental researches in electricity to a close. It began in 1831 with
his greatest discoveries, the induction of electric currents,
and the evolution of electricity from magnetism; then it
continued with terrestrial magneto-electric induction;
then with the identities of electricity from different
sources; then with conducting power generally. Then
came electro-chemical decomposition; then the electricity of the voltaic pile; then the induction of a current on itself; then static induction; then the nature
of the electric force or forces, and the character of the
electric force in the gymnotus; then the source of
power in the voltaic pile; then the electricity evolved
by friction of steam; then the magnetisation of light
and the illumination of magnetic lines of force; then
new magnetic actions, and the magnetic condition of
all matter; then the crystalline polarity of bismuth and
its relation to the magnetic form of force; then the
possible relation of gravity to electricity; then the
magnetic and diamagnetic condition of bodies, including
oxygen and nitrogen; then atmospheric magnetism;
then the lines of magnetic force, and the employment
of induced magneto-electric currents as their test and
measure; and lastly the constancy of differential magne-
crystallic force in different media, the action of heat
on magne-crystals, and the effect of heat upon the
absolute magnetic force of bodies.

The record of this work, which he has left in his manuscripts and republished in his three volumes of 'Electrical Researches,' from the papers in the 'Philosophical Transactions,' will ever remain as his noblest monument—full of genius in the conception—full of finished and most accurate work in execution—in quantity so vast that it seems impossible one man could have done so much; and this amount of work appeared still more remarkable to those who knew that Anderson's help might be summed up in two words—blind obedience.

The use of magneto-electricity in induction machines, in electrotyping, and in lighthouses, are the most important practical applications of the 'Experimental Researches in Electricity;' but it is vain to attempt to measure the stimulus and the assistance which these researches have given, and will give, to other investigators.

Lastly, the circumstances under which this work was done were those of penury. During a great part of these twenty-six years the Royal Institution was kept alive by the lectures which Faraday gave for it. 'We were living,' as he once said to the managers, 'on the parings of our own skin.' He noted even the expenditure of the farthings in research and apparatus. He had no grant from the Royal Society, and throughout almost the whole of this time the fixed income which the Institution could afford to give him was 100*l.* a year, to which the Fullerian professorship added nearly 100*l.* more.

By the 'Experimental Researches in Electricity,' Faraday's scientific life may be divided into three parts. The first, or preparatory period, lasted to 1830, when

he was thirty-nine; the second, or 'research period,' 1855.
lasted to 1855, when he was sixty-four; and the third Æt.63–64.
period of decline began in 1856 and continued to his
last report to the Trinity House in 1865. His scientific
work was carried on for fifty-two years. Out of these
the 'Experimental Researches in Electricity' occupied
more or less of twenty-six years.

The following letter from Professor Reiss, of Berlin,
the greatest statical electrician in Europe, shows how
the importance of Faraday's 'Researches in Electricity'
was recognised abroad.

PROFESSOR REISS TO FARADAY.

'Berlin : August 9, 1855.

'My dear Sir,—Returning from a journey in Silesia,
I had yesterday the great pleasure to find, as a present
from you, the third volume of the " Experimental Researches." What a wonderful work these researches
are in every respect! Incomparable for exhibiting the
greatest progresses for which science ever was indebted
to the genius of a single philosopher, highly instructive
by indicating the means whereby the great results were
found.

'If Newton, not quite without reason, has been compared to a man who ascends to the top of a building by
the help of a ladder, and cuts away most of the steps
after he has done with them, it must be said that you
have left to the follower, with scrupulous fidelity,
the ladder in the same state as you have made use
of it.

'Accept my warmest thanks for your great kindness,

to have laid in my hands the object of my continual
study and admiration.

'And believe me, dear Sir, ever to be yours most
faithfully, 'P. REISS.'

I.

The laboratory work in 1855 was again on magne-
crystallic force. The action of magnetic bodies in
different media, and at different temperatures, was the
subject of the thirtieth series of ' Experimental Researches in Electricity.' In this paper he uses his lines
of force ' as a true, searching, and as yet never-failing
representative of the one form of power possessed by
paramagnets, diamagnets, and electric currents.' In any
view of the cause of magnetic action, the results
(obtained by experiment) are true, and must therefore
be valuable. To a friend he writes : ' My recent labour
has not been very productive, and yet it is an aid to
magnetic science, and indeed a very curious one, only
its curiosity and interest will not appear so much now
as hereafter.'

This was the last of the papers on electricity which
he sent to the Royal Society, although he still worked
on in the hope of further discovery. In August, September and October, his note-book shows that time in
relation to magnetic force was the subject of research.
He also thoroughly examined Ruhmkorff's induction
apparatus, and at the end of the year he again made
experiments on the relations of light and magnetism,
but he obtained only negative results.

In the ' Philosophical Magazine ' for June there was
the translation of a paper by Dr. P. Reiss, of Berlin, on
Faraday's views regarding the action of non-conducting

bodies in electric induction. To this, Faraday sent a reply in November, and this was published with the answer of Dr. P. Reiss, in the 'Philosophical Magazine' for January 1856. Faraday added some foot-notes, and says : 'I trust they will be received, not as exciting discussion about hypothesis, but simply in explication (to the reader) of my own view. It is not the duty or place of a philosopher to dictate belief, and all hypothesis is more or less matter of belief; he has but to give his facts, and his conclusions, and so much of the logic which connects the former with the latter as he may think necessary, and then to commit the whole to the scientific world for present, and, as he may sometimes without presumption believe, for future judgment.'

For the Royal Institution, his first Friday evening discourse was on some points of magnetic philosophy, and on gravity. This was a popular view of the paper which he had sent at the end of the previous year to the 'Philosophical Magazine.'

A correspondence which took place in consequence of this lecture is of some interest.

PROFESSOR AIRY TO REV. JOHN BARLOW.

'February 7, 1855.

' My dear Sir,—You called my attention to Faraday's paper about lines of force; in some measure I think, to ask my opinion on the question therein treated.

' The following may be taken as nearly expounding my present views :—

' 1. It seems to me that the question ought to be split into two, namely, (a) Is there any reason for treating the influences of magnetism in any way different from the way of treating the effects of gravitation, &c.? (*b*) Are these influences to be considered as influences related to space, or related to the bodies sustaining their action?

'2. On question (*a*) I give my opinion without misgiving, as regards the mechanical effects. The effect of a magnet upon another magnet may be represented *perfectly* by supposing that certain parts act just as if they pulled by a string, and that certain other parts act just as if they pushed with a stick. And the representation is not vague, but is a matter of strict numerical calculation; and when this calculation is made on the simple law of the inverse square of distance, it does (numerically) represent the phenomena with precision. I can answer for this, because we are perpetually making this very calculation. I know the difficulty of predicating the effects of evidence on other people's minds, but I declare that I can hardly imagine anyone who practically and numerically knows this agreement, to hesitate an instant in the choice between this simple and precise action, on the one hand, and anything so vague and varying as lines of force, on the other hand.

'You know the French mathematicians have calculated the effect of induction accurately on the same laws.

'3. On the metaphysical question (*b*) I have only one remark to make. I do not think Faraday's remark on the bringing a new body into space is pertinent, because no new body is brought into space. We all start with the notion that the quantity of the mysterious ὑπόστασις is never altered. Therefore, when I contemplate gravitation, I contemplate it as a relation

between two particles, and not as a relation between
one particle (called the attracting particle) and the space
in which the other (called the attracted particle) finds
itself for the moment.   I contemplate it as a relation
between two particles, which relation (mechanically
considered) has respect to different directions, and has
varying magnitude : the said direction and magnitude
having very simple relations with the relative direction
and magnitude of the two particles.   I can easily conceive that there are plenty of bodies about us not subject to this intermutual action, and therefore not
subject to the law of gravitation.

'I dare say that Faraday will not be offended with this.

'I am, my dear Sir, yours very truly,

'G. B. AIRY.'

PROFESSOR AIRY TO REV. JOHN BARLOW.

'February 26, 1855.

'My dear Sir,—I have been obliged sometimes to
explain that since the reign of good king Rowland Hill
began, one idea per letter is my tariff, and request you
to understand this on the present occasion.   Moreover,
in this instance, said idea is only a supplementary idea.
It is this : in writing on Faraday's philosophy, I said
that I contemplated gravitation not as a relation between
an attracting body and space, but as a relation between
two attracting bodies ; but I omitted to point out that
this view appears to me to be in some measure established by the fact that a body which attracts is *ipso
facto* attracted according to the same law.   The land
and water of the earth attract the moon ; but the moon
also attracts the land and water, and produces tides

and precession. The earth's attraction on the moon
diminishes when the moon is in apogee; so does
the moon's attraction on the earth; and produces
small tides.

.   .   .   .   .   .

'I am, my dear Sir, yours very truly,

'G. B. AIRY.'

FARADAY TO REV. JOHN BARLOW.

'Royal Institution: February 28, 1855.

'My dear Barlow,—I return you Airy's second note.
I think he must be involved in some mystery about my
views and papers; at all events, his notes mystify me.
In the first, he splits the question into (*a*) action inversely as the square of the distance, and (*b*) meta-
physics. What the first has to do with my consideration, I cannot make out. I do not deny the law of
action referred to in all like cases; nor is there any difference as to the mathematical results (at least, if I understand Thomson and Van Rees), whether he takes
the results according to my view or that of the French
mathematicians. Why, then, talk about the inverse
square of the distance? I had to warn my audience
against the sound of this law and its supposed opposition on my Friday evening, and Airy's note shows that
the warning was needful. I suppose all magneticians
who admit differences in what is called magnetic satu-
ration in different bodies, will also admit that there
may be and are cases in which the law of the inverse
square of the distance may not apply to magnetic
action; but such cases are entirely out of the present
consideration.

'As to the metaphysical question, as it is called. If 1855. the admitted theory of gravitation will not permit us to Æt. 63 suppose a new body brought into space, so that we may contemplate its effects, I think it must be but a poor theory ; but I do not want a new body for my speculations, for, as I have said in the Friday evening paper, the motions of either planet or comet in an ellipse is sufficient base for the strict philosophical reasoning ; and if the theory will not permit us to ask a question about the conservation of force, then I think it must be very weak in its legs. The matter in the second note is quite in accordance with my views *as far as it goes*, only there is at the end of it a question which arises, and remains unanswered : When the attractive forces of the earth and moon in respect of each other diminish, what becomes of them, i.e. of the portions which disappear ?

'Ever, my dear Barlow, yours truly,

'M. FARADAY.'

PROFESSOR AIRY TO REV. JOHN BARLOW.

'Royal Observatory, Greenwich : March 3, 1855.

'My dear Sir,—I am much obliged for the sight of Faraday's note, which I have carefully read, and which I now return to you, but without comment. For what sayeth Ulysses in Pope's "Homer"? (at least said so more than forty years ago, the last time that I had an opportunity of learning his sentiments in English) :

Shall I with brave Laodamus contend ?
A friend is sacred, and I style him friend.

I think that my two notes have put you in possession

of my thoughts on the question, and that is all that I
desire.

'Yours, my dear Sir, very truly,

'G. B. AIRY.'

For the Institution he gave two more Friday evening
lectures; one on the experiments he had made on
electric conduction, which he described the previous
year in a letter to Professor de la Rive. At the end
of this lecture he said: 'But we must not dogmatize
on natural actions, or decide upon their physical nature
without proof; and, indeed, the two modes of electric
action, the electrolytic and the static, are so different,
yet each so important (the one doing all by quantity at
very low intensities; the other all by intensity, without
scarcely any proportionate quantity), that it would be
dangerous to deny too hastily the conduction proper to
a few cases in static induction, whilst it is known to be
essential to the many only because electrolytic conduction is essential to electrolytic action.'

He gave the Christmas Lectures on the metals.

For the Trinity House he only went to Birmingham
to examine some apparatus at Chance's glass works.

II.

He was made Honorary Member of the Imperial
Society of Naturalists, Moscow; Corresponding Associate
of the Imperial Institute of Sciences of Lombardy.

This year, on the application of his friend M. Dumas,
he was made Commander of the Legion of Honour,
and received the Grand Medal of Honour of the French
Exhibition for his discoveries. Early in the next year

Faraday wrote the following letter to the Emperor, and enclosed it to M. de Persigny, the French Ambassador in England.

FARADAY TO HIS EXCELLENCY THE COUNT F. DE PERSIGNY.

'Royal Institution : January 19, 1856.

' M. le Count,—I am led to believe that I ought to thank the Emperor personally for the high honour he has done me in creating me a Commander of the Legion of Honour, especially when I call to remembrance circumstances of personal communication in former times.

' May I beg the favour of the conveyance of the enclosed to its high destination.

' I have the honour to remain, your Excellency's most humble, obedient servant,

' M. FARADAY.'

FARADAY TO HIS IMPERIAL MAJESTY THE EMPEROR.

'Royal Institution: January 19, 1856.

' Sire,—I fear to intrude, yet I also fear to seem ungrateful ; and before your Majesty I would rather risk the former than the latter. I know not how to return fit thanks for the high and most unexpected honour which your Imperial Majesty has conferred upon me in the gift of the Degree of Commandant of the Legion of Honour. I cannot promise to deserve it by the future, for the effects of time tell me there are no hopes that I should hereafter work for science as in past years. I can only offer a most grateful and unfailing remembrance of that which to me is more than honour—of the kindness of your Imperial Majesty to

1856.
Æt. 64.

one such as I am ; and I feel deeply affected by the thought that even I, by your Majesty's favour, form one link, though a very small one, in the bands which I hope will ever unite France and England.

'Hoping and believing that your Majesty will accept my earnest thanks and deep-seated wishes for your Majesty in all things, I venture to sign myself as

'Your Imperial Majesty's most humble and most grateful servant,   'M. FARADAY.'

M. Dumas heard in April that Faraday had not received the insignia of Commander of the Order of the Legion of Honour. He asked Faraday to send him a note addressed to the Grand Chancellor, to ask for the insignia, which, in consequence of his absence, he was prevented from receiving from the hands of the Emperor.

Faraday in consequence wrote to the Grand Chancellor of the Legion of Honour :—

FARADAY TO THE CHANCELLOR OF THE LEGION OF HONOUR.

'London : April 28, 1856.

'Monseignor,—Though feeling quite unworthy of the high distinction done me by the Emperor when he deigned to confer upon me the degree of Commander of the Legion of Honour, I am still unwilling to resign any part of that distinction. I was in the country because of ill health, and therefore unable to be at Paris at the time when His Majesty distributed the marks of his pleasure ; but being encouraged by my scientific friends, I venture to apply to your Excellency for the insignia of the degree, and hope that the estimation in which I hold the honour may be in some measure an excuse for the liberty I am taking.

' I have the honour to be your Excellency's most humble and obedient servant,

'M. FARADAY.'

On May 13 the cross and collar were sent to M. Dumas, with an explanation that in consequence of Mr. Faraday's absence, the collar intended for him had been placed by the Emperor round the neck of M. Delacroix, the great painter. M. Dumas wrote to Faraday :—' J'estime que M. Delacroix est bien heureux de porter quelque chose qui vous appartenait.'

III.

Several letters in 1855 also show his nature, his thoughts, and his character. The most remarkable of these was published in the ' Times ' of July 9, on the state of the river Thames.

To Professor Schönbein he writes of his own state of health.

FARADAY TO PROFESSOR SCHÖNBEIN.

'Hastings: April 6, 1855.

' My dear Friend,—I have brought your letter here, that I might answer its great kindness at some time when I could remember quietly all the pleasure I have had since the time I first knew you. I say remember it all, but that I cannot do ; for as a fresh incident creeps dimly into view, I lose sight of the old ones, and I cannot tell how many are forgotten altogether. But think kindly of your old friend; you know it is not willingly, but of natural necessity, that his impressions fade away. I cannot tell what sort of a portrait you have made of me;[1] all I can say is, that whatever it may be I doubt whether I should be able to remember it; indeed, I may say, I know I should not, for I have just been under the sculptor's hands, and I look at the clay, and I look at the marble, and I look in the glass, and the more I look the less I know about the matter, and the more uncertain I become. But it is of no great consequence; label the marble, and it will do just as well as if it were like. The imperishable marble of your book will surely flatter.

'You describe your state as a very happy one— healthy, idle, and comfortable. Is it indeed so? or are you laying up thoughts which are to spring out into a rich harvest of intellectual produce? I cannot imagine you a *do-nothing*, as I am. Your very idleness must be activity. As for your book, it makes me mad to think I shall lose it. There was the other (which the " Athenæum " or some other periodical reviewed) in German, but we never saw it in English. I often lent it to others, and heard expressions of their enjoyment, and sometimes had snatches out of it, but to me it was a shut book. How often have I desired to learn German, but headache and giddiness have stopped it.

'I feel as if I had pretty well worked out my stock of original matter, and have power to do little more than reconsider the old thoughts. A friend of mine will in the course of a month or two, put the paper I speak of (in the " Philosophical Magazine ") in your way. You will therein perceive that I am as strong as ever

[1] In a German book.

in the matter of lines of magnetic force and a magnetic medium ; and, what is more, I think that men are beginning to look more closely to the matter than they have done heretofore, and find it a more serious affair than they expected. My own convictions and expectations increase continually ; *that*, you will say, is because I become more and more familiar with the idea.   It may be so, and in some manner *must* be so ; but I always tried to be very critical on myself before I gave anybody else the opportunity, and even now I think I could say much stronger things against my notions than anybody else has. Still the old views are so utterly untenable *as a whole*, that I am clear they must be wrong, whatever is right.

' Our kindest remembrances also to Mrs. Schönbein, and the favourable family critics. I can just imagine them hearing you read your MS., and flattering you up, and then giving you a sly mischievous mental poke in the ribs, &c. They cannot think better of you than I do.

' Ever, my dear Schönbein, your attached friend,

' M. FARADAY.'

Mr. W. Cox wrote to Faraday :—

' Sir,—I have staying here with me Mr. Home, who is a medium for spiritual demonstrations, and shall be very happy to give you the opportunity to show tables and chairs moving, and other phenomena much more extraordinary, without *any person* being NEAR.'

Faraday answered :—

' Royal Institution.

' Mr. Faraday is much obliged to Mr. Cox, but he will not trouble him.   Mr. Faraday has lost too much time about such matters already.'

Mr. Cox replied :—

     ' Sir,—You are wrong in not seeing me. I have facts
which are at your service NOW. After to-day they
will belong to others.

' Respectfully yours,

' W. Cox.'

FARADAY TO THE EDITOR OF THE ' TIMES.'

' Royal Institution : July 7, 1855.

     ' Sir,—I traversed this day by steamboat the space
between London and Hungerford Bridges, between
half-past one and two o'clock. It was low water, and
I think the tide must have been near the turn. The
appearance and smell of the water forced themselves at
once on my attention. The whole of the river was an
opaque pale brown fluid. In order to test the degree
of opacity, I tore up some white cards into pieces, and
then moistened them, so as to make them sink easily
below the surface, and then dropped some of these
pieces into the water at every pier the boat came to.
Before they had sunk an inch below the surface they
were undistinguishable, though the sun shone brightly
at the time, and when the pieces fell edgeways the
lower part was hidden from sight before the upper part
was under water.

     ' This happened at St. Paul's Wharf, Blackfriars
Bridge, Temple Wharf, Southwark Bridge, and Hungerford, and I have no doubt would have occurred further
up and down the river. Near the bridges the feculence
rolled up in clouds so dense that they were visible at
the surface even in water of this kind.

     ' The smell was very bad, and common to the whole
of the water. It was the same as that which now

comes up from the gully holes in the streets. The
whole river was for the time a real sewer. Having
just returned from the country air, I was perhaps more
affected by it than others ; but I do not think that I
could have gone on to Lambeth or Chelsea, and I was
glad to enter the streets for an atmosphere which,
except near the sink-holes, I found much sweeter than
on the river.

'I have thought it a duty to record these facts, that
they may be brought to the attention of those who
exercise power, or have responsibility in relation to the
condition of our river. There is nothing figurative in
the words I have employed, or any approach to exag-
geration. They are the simple truth.

' If there be sufficient authority to remove a putre-
scent pond from the neighbourhood of a few simple
dwellings, surely the river which flows for so many
miles through London ought not to be allowed to
become a fermenting sewer. The condition in which I
saw the Thames may perhaps be considered as excep-
tional, but it ought to be an impossible state ; instead
of which, I fear it is rapidly becoming the general
condition. If we neglect this subject, we cannot expect
to do so with impunity ; nor ought we to be surprised
if, ere many years are over, a season give us sad proof
of the folly of our carelessness.

' I am, Sir, your obedient servant,
' M. FARADAY.'

The Admiralty requested his opinion regarding
Crews's patent disinfecting powder, and Crews's anti-
miasma lamp, to be used in ships and hospitals. He
replied to Thomas Phinn, Esq., M.P. :—

FARADAY TO THE SECRETARY OF THE ADMIRALTY.

'Royal Institution : August 27, 1855.

'Sir,—I am sure that when the Lords Commissioners of the Admiralty look again at the enclosed printed advertising paper which you have sent to me, and which I return herewith, they will see that it is not such a document as I can be expected to give an opinion upon. My Lords will do me the favour to remember that, as I have said on former occasions, though I am always willing to help the Government in important cases, and when it is thought that others cannot give satisfactory information, still I am not professional ; and being engaged in deep philosophic research, am desirous of having my time and thoughts as little engaged by extraneous matters as possible.

'I have the honour to be, Sir, your very obedient servant,   'M. FARADAY.'

To Professor Matteucci he wrote his views regarding the lines of force and Tyndall's work on the relations of paramagnetic and diamagnetic bodies.

FARADAY TO PROFESSOR MATTEUCCI.

'November 2, 1855.

'My dear Matteucci,—When I received your last, of October 23, I knew that Tyndall would return from the country in a day or two, and so waited until he came. I had before that told him of your desire to have a copy of his paper, and I think he said he would send it to you ; I have always concluded he did so, and therefore thought it best to continue the same open practice and show him your last letter, note and all. As I expected, he expressed himself greatly obliged by your consideration, and I have no doubt will think on, and repeat, your form of experiment; but he wished you to have no difficulty on his account. I conclude he is quite assured in his own mind, but does not for a moment object to counter views, or to their publication: and I think feels a little annoyed that you should *imagine for a moment* that he would object to or be embarrassed by your publication. I think in that respect he is of my mind, that we are all liable to error, but that we love the truth, and speak only what at the time we think to be truth; and ought not to take offence when proved to be in error, since the error is not intentional; but be a little humbled, and so turn the correction of the error to good account. I cannot help thinking that there are many apparent differences amongst us, which are not differences in reality. I differ from Tyndall a good deal in phrases, but when I talk with him I do not find that we differ in facts. That phrase *polarity* in its present undefined state is a great mystifier (3307, 3308).[1] Well! I am content, and I suppose he is, to place our respective views before the world, and there leave them. Although often contradicted, I do not think it worth while reiterating the expressions once set forth; or altering them, until I either see myself in the wrong or misrepresented; and even in the latter case, I let many a misrepresentation pass. Time will do justice in all these cases.

'One of your letters asks me, "What do you conceive the nature of the lines of magnetic force to be?" I think it wise not to answer that question by an assumption, and therefore have no further account to give of such physical lines than that is already given in my various papers. See that referred to already in the

[1] 'Philosophical Magazine,' February 1855.

" Philosophical Magazine " (3301–3305); and I would ask you to read also 3299, the last paragraph in a paper in the "Philosophical Magazine," June 1852, which expresses truly my present state of mind.

'But a physical line of force may be dealt with experimentally, without our *knowing its intimate physical nature*. A ray of light is a physical line of force; it can be proved to be such by experiments made whilst it was thought to be an *emission*, and also by other experiments made since it has been thought to be an *undulation*. Its physical character is not *proved* either by the one view or the other (one of which must be, and both may be wrong), but it is proved by the *time* it takes in propagation, and by its curvatures, inflexions, and physical affections. So with other physical lines of force, as the electric current; we know no more of the physical nature of the electric lines of force than we do of the magnetic lines of force; we fancy, and we form hypotheses, but unless these hypotheses are considered equally likely to be false as true, we had better not form them; and therefore I go with Newton when he speaks of the *physical lines of gravitating force* (3305 note), and leave that part of the subject for the consideration of my readers.

'The use of *lines of magnetic force* (without the *physical*) as true representations of nature, is to me delightful, and as yet never failing; and so long as I can read your facts and those of Tyndall, Weber, and others by them, and find they all come into one harmonious whole, without any contradiction, I am content to let the erroneous expressions, by which they *seem* to differ, pass unnoticed. It is only when a fact appears that *they cannot* represent that I feel urged to examina-

tion, though that has *not yet* happened.   All Tyndall's
results are to me simple consequences of the tendency
of paramagnetic bodies to go from weaker to stronger
places of action, and of diamagnetic bodies to go from
stronger to weaker places of action, combined with the
true polarity or direction of the lines of force in the
places of action.   And this reminds me of a case you
put in one of your letters, which to me presents no
difficulty :—" *a piece of bismuth on which the pole p*

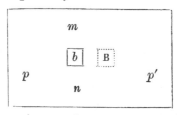

acts suffers an action on the part of the pole p', which is
the same as if the pole p' did not act or was a pole of
the contrary name."   *p*, being an S pole, repels *b*, and
sends it from a stronger into a weaker part of the field,
i.e. from $\boxed{b}$ to $\boxed{\mathrm{B}}$; then *p'* being brought up, and being
also an S pole, B is no longer the weaker place of action
but *b* ; and hence the bismuth goes back.   And that
it is the weaker place of action can be shown by a
minute magnetic needle or a crystal of bismuth, and in
many other ways (3341, &c., especially 3350).   But
suppose *p'* is selected, an N pole, then the lines of force
between *p'* and *p* are greatly strengthened in power,
and the small needle, or crystal bismuth, shows it to be
so ; but still B is no longer a weaker place of power
than *b*, and if the bismuth can only move along the line
p p' it must move from B to *b*, for *b* is the weakest
place of action in that line ; but this is a place of
unstable equilibrium, and, as you know very well, if it

can move in the line *m n*, it will move either towards *m* or towards *n*, as it happens to be on one or the other side of the axial line of the magnetic field.

'These principles, or rather laws, explain to me all those movements obtained by Tyndall against which your note is directed, and therefore I do not see in his experiments any proofs of a defined or inverse polarity in bismuth, beyond what we had before. He has worked out *well* the antithetical relations of paramagnetic and diamagnetic bodies ; and distinguished mixed actions, which by some have been much confused; but the true nature of polarity, and whether it is the same, or reversed in the two classes, is to my mind not touched. What a quantity I have written to you, all of which has no doubt been in your own mind, and tried by your judgment. Forgive me for intruding it. Ever truly yours,

'M. FARADAY.

'I am sorry to refer you to the "Philosophical Magazine." I have a third volume of my "Experimental Researches" on my desk waiting for you; it contains them all. I have not yet found a means of sending it.'

Another affectionate letter to Professor Schönbein must be preserved here.

FARADAY TO PROFESSOR SCHÖNBEIN.

'Royal Institution : November 6, 1855.

'My dear Schönbein,—It is quite time I should write you a letter, even though I may have nothing to say, and yet I surely have something to write, though it may not be philosophy, for I trust affection will last out philosophy ; and indeed were it not so I should fear that I was indeed becoming a worn out, worthless thing. But your last letter abounded in *all* matter, both the *philosophical* and also the *domestic* and *kind*, and I thank you heartily for it. That one day in the country ! How I wish I had been with you ; but I could not now walk in Switzerland as I have done in former years. All things suffer a change. May your changes be long deferred, for you must be very happy as you are ; and so am I, but my happiness is of a quieter kind than it used to be, and probably more becomes a man sixty-four years of age. And as we, i.e. my wife and I, go on our way together, our happiness arises from the same things, and we enjoy it together with I hope thankfulness to the Giver of every good and perfect gift.

.   .   .   .   .   .

' Your accounts and observations are most interesting and exciting, but I dare not try to pursue the subject, for even the matter I have in magnetism is often too much for me, and I am obliged to lay it by for a while, so that I am forbidden by nature to take up any new series of thought. But that ozone, that oxygen, which makes up more than half the weight of the world, what a wonderful thing it is, and yet I think we are only at the beginning of the knowledge of its wonders.

' My very kind remembrances to M. Wiedeman also. It is delightful to see thinking-workers rise up in science.

' Believe me to be, my dear Schönbein, your faithful friend,   ' M. FARADAY.'

CHAPTER IV.

HAMPTON   COURT—USE   OF   THE   ELECTRIC   LIGHT   IN   LIGHT-
HOUSES—DECLINE AND END OF HIS LIFE.

1856.   EVEN in the last chapter of Faraday's life each one of
ÆT.64–65. his great qualities can be very distinctly traced.

Few of those who saw him enjoying the kindness
which gave him his house at Hampton Court, or
delighting in the beauty of the sunsets from the palace
gardens, or rejoicing in the idleness of the summer life
in the country, knew that during a great part of this
period of his life he was proving by experiment whether
his magneto-electric light could be made by Professor
Holmes practically useful for lighthouses.

His energy and truthfulness made him take the
whole responsibility of the decision upon himself, and
without doubt his frequent journeys to the South
Foreland and Dungeness lighthouses, and his night
excursions in the Channel during the winter, when he
was seventy years old, were remote causes of his last
illness.

Throughout all the reports which he made regarding
the light, there is scarcely a word to show that he ever
thought of it as his light, his greatest discovery; he
even heard others call it their light without making a
remark ; but he gave all credit and honour to him who
applied it, and only said of himself, that he must take

care ' that we do not lead our authorities into error by the advice given.'

Another remarkable instance of his humility may be seen at the end of the chapter, in his burial. He knew full well that he had earned his monument in the consecrated palace of the dead, but he ' desired a gravestone of the most ordinary kind in the simplest earthly place ;' the unconsecrated ground he thought good enough to be his grave, and the silent service at his funeral consisted only of the tears and thoughts of the few relations who he wished to have there.

Although the ' Researches in Electricity ' had come to an end, yet the work in the laboratory in 1856, as seen in his note-book, was more continuous than ever before. The subject was the experimental relations of gold and other metals to light. In the laboratory book, nearly three hundred folio pages are filled with the account of his experiments which were continued throughout the whole year. His paper was sent to the Royal Society in November, and was read as the Bakerian lecture in February 1857.

In the beginning, he says he looked at the subject of light as standing between the coarser mechanical actions of matter, and the action of force at a distance, and ' admitting for the time the existence of the ether, I have often struggled to perceive how far that medium might account for or mingle in with such actions generally, and to what extent experimental trials might be devised which, with their results and consequences, might contradict, confirm, enlarge, or modify the idea we form of it ; always with the hope that the corrected or instructed idea would approach more and more to the truth of nature, and in the fulness of time coincide with it.'

'At one time I had hoped that I had altered one coloured ray into another by means of gold, which would have been equivalent to a change in the number of undulations; and though I have not confirmed that result as yet, still those I have obtained seem to me to present a useful experimental entrance into certain physical investigations respecting the nature and action of a ray of light. I do not pretend that they are of great value in their present state, but they are very suggestive, and they may save much trouble to any experimentalist inclined to pursue and extend this line of investigation.'

For the Institution he gave two Friday discourses; the first on certain magnetic actions and affections, and the second on M. Petitjean's process for silvering glass and some observations on divided gold. He again gave the Christmas Lectures on attractive forces.

He made five reports to the Trinity House, and he entered into an engagement to give advice regarding the Board of Trade lighthouses, and made four reports. Two were on Cape Race lighthouse, and one on Dr. Normandy's distilled water apparatus.

He was made Corresponding Member of the Netherland Society of Sciences, Batavia; and Member of the Imperial Royal Institute of Padua.

The letters of this year that remain show the kindness of his nature, and the greatness of his reputation. He spoke more strongly than ever before of his failing memory to his friends Professors de la Rive and Schönbein, and he was much pleased with the following letters which he received when his Juvenile Lectures were finished.

H.R.H. THE PRINCE OF WALES TO FARADAY.

'Windsor Castle : January 16, 1856.

' Dear Sir,—I am anxious to thank you for the advantage I have derived from attending your most interesting lectures. Their subject, I know very well, is of great importance, and I hope to follow the advice you gave us of pursuing it beyond the lecture-room ; and I can assure you that I shall always cherish with great pleasure the recollection of having been assisted in my early studies in chemistry by so distinguished a man.

'Believe me, dear Sir, yours truly,

'ALBERT EDWARD.'

H.R.H. THE PRINCE ALFRED TO FARADAY.

'Windsor Castle : January 16, 1856.

' Dear Sir,—I write to thank you very much for the pleasure you have given me by your lectures, and I cannot help hoping they will not be the last I shall hear from you. Their subject was very interesting, and your clear explanations made it doubly so.

'Believe me, dear Sir, yours truly,

'ALFRED.'

FARADAY TO PROFESSOR SCHÖNBEIN.

'Royal Institution : March 21, 1856.

' My dear Schönbein,—The heartiest and the kindest wishes to you and the best thanks for your letter. I have it not here (Norwood, for I am resting a head like a sieve), but I know it was very pleasant, and I think contained some family details which made me long to be with you; but the fact is, that when I am with a friend, I soon need to get away again, because of the labour of conversation, and its strain upon recollection.

'I do not recollect any news, and shall be glad to lay my head down again; so with kindest remembrances. . . .

'I remain as ever, your affectionate friend,

'M. FARADAY.'

FARADAY TO PROF. A. DE LA RIVE.

'Royal Institution: March 21, 1856.

'My dear de la Rive,—Though unable to write much, I cannot longer refrain from acknowledging your kindness in sending me such a remembrance of you as the vol. ii., and in giving utterance to the great delight with which I have read it. I rejoice to think that such a work should be reprinted in the English language, for now, when asked for a good book on electricity, I know what to say.

'I will not say that I envy you for your wonderful stores of knowledge regarding all that concerns our beloved science, but I cannot help contrasting your power with mine, and wishing for a little of the ability by which a mind such as yours calls up to present remembrance what it had found worthy to lay up in its treasury.

'But we both have reasons of a higher nature than any that science can afford, to be thankful for that we have received, and not to forget the many benefits bestowed upon us, and I hope that I am not envious of you or of any man, but would rather rejoice in your exaltation.

'With the kindest remembrance of Madame de la 　1856.
Rive and of yourself, 　　　　　　　　　　　　　　　Æt. 64.

'I am, my dear friend, most truly yours,
' M. FARADAY.'

M. PLÜCKER TO FARADAY.

'Bonn: March 24, 1856.

'Dear Sir,—Looking at the date of your last kind letter, I am very much surprised my silence has been so long a one. Being another time, by election, at the head of the University, I am for a year rather entirely distracted from scientific working; therefore, that I may not fall into my former indolence, I write to you the very first day of "vacancies."'

．　　　．　　　．　　　．　　　．　　　．
．　　　．　　　．　　　．　　　．　　　．

And then he continues :—

'Dear Sir,—The extreme kindness with which you received my very first experimental researches is that moment in my scientific life at which I look back with the greatest satisfaction. When recently I had the honour to be elected a Foreign Member of the Royal Society, the origin of it is to be found only in that kindness.

'With all my heart, and for ever, yours,
' PLÜCKER.'

FARADAY TO PROFESSOR SCHÖNBEIN.

'Royal Institution: April 11, 1856.

'My dear Schönbein,—

．　　　．　　　．　　　．　　　．　　　．

'Most hearty thanks for your very pleasant, interesting picture of juvenile life. I could have enjoyed it very

much indeed. I suppose you were about the biggest
child there.

.   .   .   .   .

'I trust you will soon have the volume—which
receive favourably for my sake.

'Ever yours,

'M. FARADAY.'

FARADAY TO PROFESSOR SCHÖNBEIN.

'Royal Institution: October 14, 1856.

'My dear Friend,—Hearty, and healthy, and occupied,
and happy as you are, let me congratulate you, for
every letter of yours brings me evidence of the existence
of a healthy mind in a sound body. How you have
been running about! and you go home as if you were
refreshed rather than tired by it. I do not feel so any
longer; even if I go away for a little general health,
I am glad to return home again for rest in the company
of my dear wife and niece; but, as the Wise man hath
said, there is a time for all things, and my time is to be
quiet and look on, which I am able to do with great
content and satisfaction.

.   .   .   .   .   .

'What you tell me of your paper makes me long to
hear the whole of it, though the very pleasure of
getting knowledge is now mingled with some thoughts
of regret at the consciousness that I may quickly
lose it again. Well, a time for all things. I have been
occupying myself with gold this summer; I did not
feel head-strong enough for stronger things. The
work has been of the mountain and mouse fashion, and
if I ever publish it, and it comes to your sight, I dare

say you will think so. The transparency of gold, its division, its action on light, &c. &c. &c.

'Ever, my dear Schönbein, yours most truly,

'M. FARADAY.'

L. AGASSIZ TO FARADAY.

'Cambridge, U. S.

'My dear Sir,—It is so long since I have had the pleasure of direct intercourse with you that I might apprehend you would have forgotten me, were there not in my past recollections such circumstances as insure for me a place in your memory, I hope. You have surely not cast from your mind the enthusiastic fish-man whom you met at Dr. Mantell's in Brighton seventeen years ago, and who at that time was so happy to pay homage to the great physicist in England. You were then already old in the walks of science, and, for my part, I shall never forget the impression which this contrast between celebrity and age made upon my mind, and I can hardly believe it has escaped your attention then. Though the nature of my studies has not drawn me nearer to you since, I may hope that scientific men in all departments feel sympathy for one another; and it is on that ground I take the liberty to introduce to you my friend, one of your admirers, who is now visiting again the continent of Europe, where he has studied formerly, and who wants to pay you his respects whilst in England.

'Believe me, dear Sir, with high respect, sincerely yours,

'L. AGASSIZ.'

W. THOMSON TO FARADAY.

'2 College, Glasgow.

' My dear Sir,—Although I hope soon to see you in London, I cannot delay till then thanking you for your letter and for the very kind expressions it contains. Such expressions from you would be more than a sufficient reward for anything I could ever contemplate doing in science. I feel strongly how little I have done to deserve them, but they will encourage me with a stronger motive than I have ever had before to go on endeavouring to see in the direction you have pointed, which I long ago learned to believe is the direction in which we must look for a deeper insight into nature.

' I cannot explain to you how much I fall short of deserving what you say, but must simply thank you most sincerely for your kindness in writing as you have done.

' Believe me, ever yours truly,

'WILLIAM THOMSON.'

In 1857 there was but little work in the laboratory, except in August and September. The subject of investigation was on time in magnetism. In one paragraph, he writes: ' *Time.* It would appear very hopeless to find the time in magnetic action, if it at all approached to the time of light, which is about 190,000 miles in a second, or that of electricity in copper wire, which approximates to the former. But these powers, which act on interposed media, are known to vary, and sometimes wonderfully. Thus the time of action at a distance by conduction is wonderfully different for

electricity in copper, water, and wax. Nor is it likely 1857.
that the paramagnetic body oxygen can exist in the Æt.65-66.
air and not retard the transmission of the magnetism.
At least, such is my hope.

'As to the detection, a difference of $\frac{1}{30}$th of an inch can
be seen with a radius of ten feet, and it is the $\frac{1}{22620}$th
part. Suppose we say that the light lines will be visible
with a revolution of the contact mirror thrice in a
second, that is equivalent to a revolution of the light ray
six times in a second, so that $22620 \times 6 = 135720$, so
that the space moved through in the $\frac{1}{135720}$th of a
second will probably be easily distinguished. If that
be the time for conduction through 100 feet of distance,
it would show a transmission of magnetic force with
a velocity of 135720×100, or 13,572,000 feet in a
second, or 2574 miles, or about $\frac{1}{74}$th part of that of light.
Probably, the radius may be doubled or tripled—
perhaps the rotation be much increased; but then the
difficulty will be to catch the moment of cessation, for
the impression of the preceding lines of light will
remain on the eye if the revolutions are more than ten
in a second.'

To the Rev. John Barlow he writes :—

'Highgate : August 19, 1857.

'I am in town, and at work more or less every day.
My memory wearies me greatly in working; for I cannot remember from day to day the conclusions I come
to, and all has to be thought out many times over. To
write it down gives no assistance, for what is written
down is itself forgotten. It is only by very slow
degrees that this state of mental muddiness can be
wrought either through or under; nevertheless, I know

that to work somewhat is far better than to stand still, even if nothing comes of it. It is better for the mind itself—not being quite sure whether I shall ever end the research, and yet being sure, that if in my former state of memory, I could work it out in a week or two to a successful and affirmative result.

'Do not be amazed by what I am telling you : it is simply the thing I remember to tell you. If other things occurred to my mind, I would tell you of them. But one thing which often withholds me, is, that if I begin a thing, I find I do not report it correctly, and so naturally withdraw from attempting it. One result of short memory is coming curiously into play with me. I forget how to spell. I dare say if I were to read this letter again, I should find four or five words of which I am doubtful, " withholds, wearies, successful," &c.; but I cannot stop for them, or look to a dictionary (for I had better cease to write altogether), but I just send them, with all their imperfections, knowing that you will receive them kindly.

'Ever, dear Barlow, truly yours,
'M. FARADAY.'

In a note to Professor Schönbein he alludes to his work at this time.

FARADAY TO SCHÖNBEIN.

'Royal Institution : November 24, 1857.

' My dear Schönbein,—I expected you would have seen much of your last letter in the " Philosophical Magazine " before now. . . .

' What a wonderful thing oxygen is, and so I suppose would every other element appear if our knowledge were more perfect.

'I ventured to send you a paper the other day by the
post . . . about gold, and the relation of it and other
metals to light. Many facts came out during the
inquiry which surprised me greatly; especially the
effects of pressure; and also those relating to polarized
light. Lately I have been working at the relation of
time to actions at a distance, as those actions which
class as magnetic ; but the subject is very difficult: the
requisite apparatus requires to be frequently remade,
each time being more perfected ; and whether I shall
catch the $\frac{1}{200000}$th part of a second (if required)
seems very doubtful. In the meantime, I am for the
present tired, and must lay the research on the shelf.

'I have undertaken to give half a dozen juvenile
lectures after Christmas; whether they will come off
(as we say) or not, is doubtful. Patience.

'I do not think we have much scientific news ; at
least, I do not hear of much ; but then I do not go
within reach of the waves of sound, and so must
consent to be ignorant. Indeed, too much would drive
me crazy in the attempt to hold it.

'Ever, my dear Schönbein, yours most truly,

'M. FARADAY.'

FARADAY TO C. MATTEUCCI.

'Royal Institution : March 20, 1857.

'I snatch a few weary moments to have a word with
you in the way of thanks for your letters of news.

.   .   .   .   .

'I won't pretend to send you any news, for, when I
try to remember it, all is slow to me, and you are so

active and *spirituel* that it seems to me as if you were out of my sight.

'Perhaps you may be inclined to say as regards gravity, and that I am out of your sight ; nevertheless, I have a strong trust and conviction,

'I am ever yours,

'M. FARADAY.'

On February 27 he gave a lecture for the Royal Institution on the conservation of force. It begins thus :—'Various circumstances induce me at the present moment to put forth a consideration regarding the conservation of force. . . . There is no question which lies closer to the root of all physical knowledge than that which inquires whether force can be destroyed or not. . . . Agreeing with those who admit the conservation of force to be a principle in physics as large and sure as that of the indestructibility of matter, or the invariability of gravity, I think that no particular idea of force has a right to unlimited and unqualified acceptance that does not include assent to it. . . . Supposing the truth of the principle is assented to, I come to its uses. No hypothesis should be admitted nor any assertion of a fact credited that denies the principle. . . . The received idea of gravity appears to me to ignore entirely the principle of the conservation of force ; and by the terms of its definition, if taken in an absolute sense, " *varying* inversely as the square of the distance," to be in direct opposition to it.'

Two remarkable letters in connection with this lecture have been preserved ; the one was written to the Rev. Edward Jones, and the other to Mr. Clerk Maxwell. In this last letter he says : 'I perceive that I do

not use the word " force " as you define it, " the ten-   1857.
dency of a body to pass from one place to another."   Æt.65–66.
What I mean by the word is the source or sources of
all possible actions of the particles or materials of the
universe.' His clear idea was ' that it was impossible
to create force. We may employ it, we may invoke it
in one form by its consumption in another ; we may
hide it for a period, but we can neither create nor
destroy it. We may cast it away, but where we dismiss
it there it will do its work.'

In 1831, his great discovery of magneto-electricity
came from his holding ' an opinion almost amounting
to conviction,' that as electricity could produce magnetism, therefore magnetism must produce electricity.
When working on the electricity of the voltaic pile in
1834, he held ' that chemical and electrical forces were
identical.' He wrote in his note-book, ' electricity is
chemical affinity : chemical affinity is electricity.' In
his lecture at the Royal Institution, June 21, 1834, on
the relations of chemical affinity, electricity, heat,
magnetism, and other powers of matter, he showed by
experiment ' that all are connected.' ' That the production of any one from another, or the conversion of
one into another, may be observed.' ' We cannot say
that any one is the cause of the others, but only that
all are connected and due to a common cause.'

In 1845, in his paper on the magnetism of light, he
says : ' I have long held an opinion almost amounting
to conviction, in common I believe with many other
lovers of natural knowledge, that the various forms
under which the forces of matter are made manifest
have one common origin ; or, in other words, are so
directly related and mutually dependent, that they are

convertible, as it were, one into another, and possess equivalents of power in their action.'

In February 1849 he says: 'The exertions in physical science of late years have been directed to ascertain not merely the natural powers, but the manner in which they are linked together, the universality of each in its action, and their probable *unity in one*.' ' I cannot doubt that a glorious discovery in natural knowledge, and of the wisdom and power of God in creation, is awaiting our age, and that we may not only hope to see it, but even be honoured to help in obtaining the victory over present ignorance and future knowledge.'

And in his last paper in 1860 he wrote: 'Under the full conviction that the force of gravity is related to the other forms of natural power, and is a fit subject for experiment, I endeavoured, on a former occasion—1851—to discover its relation with electricity, but unsuccessfully. Under the same deep conviction, I have recently striven to procure evidence of its connection with either electricity or heat. By a relation of forces I do not mean the production of effects associated usually with one form of power by the exercise of another, unless the results are direct. There is no difficulty in obtaining either electrical or heating phenomena indirectly from the force of gravity.'

Throughout his life, Faraday worked for this grand generalisation. But his experiments would not let him make it. The equivalent transformation of different motions was proved by experiment long after he had pointed out ' the relationship of the forces.' The constancy of the sum of potential and of actual energies is now considered as certain, and the permanency of the causes of energy is assumed to be true. But the unity

of these causes, like the existence of a single parent
elementary matter, remains a problem which experiment will perhaps never be able to establish, although
the genius of Faraday led him to consider that it was
'naturally capable of attack.'

To the Rev. Edward Jones, of West Peckham, Maid-
stone, he says :—

'Royal Institution : June 9, 1857.

'My dear Sir,—I have received your very kind letter
and paper, and am delighted at such a result of my
evening. If nothing else had come of it but that, it
would have been a sufficient reward ; but much else
has come, and I expect much more.

'I do not think you can find in my papers any word
or thought that contradicts the law of gravitating
action. My observations are all directed to the *definition or description of the force* of gravitation with the
view of clearing up the received idea of the force, so
that if inaccurate or insufficient it may not be left as
an obstacle in the present progressive state of science.

'If I am wrong in believing that according to the
present view the mutual gravitating force of two particles, A and B, remains unchanged, whatever other
particles come to bear upon A and B, then the sooner
I am corrected publicly the better.

'If your view (whether old or new), that the power
of A remains unchanged in amount, but is subdivided
upon every particle which acts upon it, is the true or
the accepted one, then I shall long to see it published
and acknowledged, for I do not find it received at
present. I have proved to my own satisfaction that

such is the case with the dual powers electricity [1] and magnetism,[2] and it is the denial of it as regards gravity which makes up my chief difficulty in accepting the established view of that power. Your statement that A may attract or act on B C with a force of one, whilst B C act on A with a force of two, seems to me inconsistent with the law that action and reaction are equal; but I suppose I am under some misconception of your meaning.

'The cases of action at a distance are becoming, in a physical point of view, daily more and more important. Sound, light, electricity, magnetism, gravitation, present them as a series.

'The nature of sound and its dependence on a medium we think we understand pretty well. The nature of light as dependent on a medium is now very largely accepted. The presence of a medium in the phenomena of electricity and magnetism becomes more and more probable daily. We employ ourselves, and I think rightly, in endeavouring to elucidate the physical exercise of these forces, or their sets of antecedents and consequents, and surely no one can find fault with the labours which eminent men have entered upon in respect of light, or into which they may enter as regards electricity and magnetism. Then what is there about gravitation that should exclude it from consideration also? Newton did not shut out the physical view, but had evidently thought deeply of it; and if he thought of it, why should not we, in these advanced days, do so too? Yet how can we do

---

[1] *Experimental Researches,* 8vo. vol. i. par. 1177, 1215, 1681, &c.

[2] *Experimental Researches,* xxviii. vol. iii. p. 328, &c., par. 3109, 3121, 3225, and also par. 3324 of the same vol. iii. p. 544.

1857.
Æt. 65.

so if the present definition of the force, as I understand it, is allowed to remain undisturbed; or how are
its inconsistencies or deficiencies as a description of the
force to be made manifest, except by such questions
and observations as those made by me, and referred to
in the last pages of your paper? I believe we ought
to search out any deficiency or inconsistency in the
sense conveyed by the received form of words, that
we may increase our real knowledge, striking out or
limiting what is vague. I believe that men of science
will be glad to do so, and will even, as regards gravity,
amend its description, if they see it is wrong. You
have, I think, done so to a large extent in your manuscript, and I trust (and know) that others have done
so also. That I may be largely wrong I am free to
admit—who can be right altogether in physical science,
which is essentially progressive and corrective. Still
if in our advance we find that a view hitherto accepted
is not sufficient for the coming development, we ought
I think (even though we risk something on our own
part), to run before and rise up difficulties, that we
may learn how to solve them truly. To leave them
untouched, hanging as dead weights upon our thoughts,
or to respect or preserve their existence whilst they
interfere with the truth of physical action, is to rest
content with darkness and to worship an idol.

'I take the liberty of sending by this post copies of
two papers. The one on conservation of force is, I
suppose, that which you have read. I have made remarks in the margin which I think will satisfy you
that I do not want to raise objections, except where the
definition of gravity originates them of itself. The
other is on the same subject two years anterior. If

you would cause your view of gravity as a force *un-changing* in amount in A, but disposable in part towards one or many other particles, to be acknowledged by scientific men; you would do a great service to science. If you would even get them to say yes or no to your conclusions, it would help to clear the future progress. I believe some hesitate because they do not like to have their thoughts disturbed. When Davy discovered potassium it annoyed persons who had just made their view of chemical science perfect; and when I discovered the magneto-electric spark, distaste of a like kind was felt towards it, even in high places. Still science must proceed; and with respect to my part in the matter of gravitation, I am content to leave it to the future. I cannot help feeling that there is ground for my observations, for if there had been an evident answer it must have appeared before now. That the answer, when it comes, may be different to what I expect, I think is very probable, but I think also it will be as different from the present received view. Then a good end will be obtained, and indeed your observations and views appear to me to be much of that kind.

'If it should be said that the physical nature of gravitation has not yet been considered, but only the law of its action, and, therefore, that no definition of gravity as a power has hitherto been necessary; that may be so with some, but then it must be high time to proceed a little further if we can, and that is just one reason for bringing the principle of the conservation of force to bear upon the subject. It cannot, I think, for a moment be supposed that we are to go no further in the investigation. Where would our knowledge of

light, or magnetism, or the voltaic current have been 1857.
under such a restraint of the mind? Æt. 66.

'Again thanking you most truly for the attention
you have given to me and the subject, I beg you to
believe that

'I am, very gratefully, your faithful servant,

'M. FARADAY.'

FARADAY TO MR. CLERK MAXWELL.

'Royal Institution: November 13, 1857.

'My dear Sir,—If on a former occasion I seemed to
ask you what you thought of my paper, it was very
wrong, for I do not think anyone should be called upon
for the expression of their thoughts before they are
prepared and wish to give them. I have often enough
to decline giving an opinion, because my mind is not
ready to come to a conclusion, or does not wish to be
committed to a view that may by further consideration
be changed. But having received your last letter, I am
exceedingly grateful to you for it; and rejoice that my
forgetfulness of having sent the former paper on conservation has brought about such a result. Your letter
is to me the first intercommunication on the subject
with one of your mode and habit of thinking. It will
do me much good, and I shall read and meditate on it
again and again.

'I dare say I have myself greatly to blame for the
vague use of expressive words. I perceive that I do
not use the word " force " as you define it, " the tendency of a body to pass from one place to another."
What I mean by the word is the source or sources of

all possible actions of the particles or materials of the universe, these being often called the powers of nature when spoken of in respect of the different manners in which their effects are shown.

'In a paper which I have received at this moment from the " Phil. Mag.," by Dr. Woods, they are called the *forces*, " such as electricity, heat, &c." In this way I have used the word " force " in the description of gravity which I have given as that expressing the received idea of its nature and source, and such of my remarks as express an opinion, or are critical, apply only to that sense of it. You may remember I speak to labourers like myself; experimentalists on force generally who receive that description of gravity as a physical truth, and believe that it expresses all and no more than all that concerns the nature and locality of the power,—to these it limits the formation of their ideas and the direction of their exertions, and to them I have endeavoured to speak, showing how such a thought, if accepted, pledged them to a very limited and probably erroneous view of the cause of the force, and to ask them to consider whether they should not look (for a time, at least), to a source in part external to the particles. I send you two or three old printed papers with lines *marked* relating to this point.

'To those who disown the definition or description as imperfect, I have nothing to urge, as there is then probably no real difference between us.

'I hang on to your words, because they are to me weighty; and where you say, " I, for my part, cannot realise your dissatisfaction with the law of gravitation, provided you conceive it according to your own principles," they give me great comfort. I have nothing to say against the law of the action of gravity. It is against the law which measures its total strength as an inherent force that I venture to oppose my opinion; and I must have expressed myself badly (though I do not find the weak point), or I should not have conveyed any other impression. All I wanted to do was to move men (not No. 1, but No. 2), from the unreserved acceptance of a principle of physical action which might be opposed to natural truth. The idea that we may possibly have to connect *repulsion* with the lines of gravitation-force (which is going far beyond anything my mind would venture on at present, except in private cogitation), shows how far we *may* have to depart from the view I oppose.

'There is one thing I would be glad to ask you When a mathematician engaged in investigating physical actions and results has arrived at his own conclusions, may they not be expressed in common language as fully, clearly, and definitely as in mathematical formulæ? If so, would it not be a great boon to such as we to express them so—translating them out of their hieroglyphics that we also might work upon them by experiment. I think it must be so, because I have always found that you could convey to me a perfectly clear idea of your conclusions, which, though they may give me no full understanding of the steps of your process, gave me the results neither above nor below the truth, and so clear in character that I can think and work from them.

'If this be possible, would it not be a good thing if mathematicians, writing on these subjects, were to give us their results in this popular useful working state as well as in that which is their own and proper to them?

> 'Ever, my dear Sir, most truly yours,
>
>   'M. FARADAY.'

His second lecture for the Institution was on the relations of gold to light. At the end of the lecture he said : 'The object of these investigations was to ascertain the varied powers of a substance acting upon light when its particles were extremely divided, to the exclusion of every other change of constitution.' In his notes he wrote : 'We are looking for the first small, and as yet unknown, effects, for these grow up into the great advanced truths of science—as a bubble—a breaking forth—a cascade—a storm—Niagara or Schaffhausen —the cannon's mouth.'

After Easter he gave a course of six lectures on static electricity.

An account of a ready method of determining the presence, position, depth, length and motion (if any) of a needle broken into the foot, was published in the 'Proceedings' of the Medical and Chirurgical Society for March 18th. In this, as in all that he did, the perfection of his hand work as well as his head work is to be seen.

He gave the Juvenile Lectures on electricity.

He made six reports to the Trinity House. The most important was on Holmes's magneto-electric light, which was put up at Blackwall, and observed from Woolwich, and compared with a Fresnel lamp in the centre of Bishop's lens, and also in the focus of a parabolic reflector. He critically examined the cost of the apparatus, the price of the light, the suppositions

regarding its intensity and advantages, and the propo-   1857.
sition to put one up in a lighthouse.   He agreed to the   Æт.65–66.
trial of the light at the South Foreland.

He was made Member of the Institute of Breslau ;
Corresponding Associate of the Institute of Sciences,
Venice; and Member of the Imperial Academy, Breslau.

As evidences of his reputation, a letter to Dr.
Percy, regarding the Presidentship of the Royal Society,
and from Professor Hansteen of Christiania, are of
interest; and to his niece, Miss Reid, his religious
nature is shown in words which he seldom uttered.

A deputation from the council of the Royal Society,
consisting of the President (Lord Wrottesley), Mr.
Grove, and Mr. Gassiot, went to Faraday to urge his
acceptance of the Presidentship.   Dr. Percy, who was
one of the council, wrote privately to him.

He answered :—

FARADAY TO DR. PERCY.

'Royal Institution : May 21, 1857.

'My dear Percy,—Your letter is very kind and
earnest, and I thank you heartily for it, but I may not
change my conclusion.   None can know but myself
how unfit it would be.

'Ever affectionately yours,
'M. FARADAY.'

PROFESSOR HANSTEEN TO FARADAY.

'Observatory, Christiania : December 30, 1857.

'Dear and honoured Sir,—I thank you heartily for
your letter of December 16.   At first, while you have
written yourself, as you could better declare the circumstances ; and secondly, while I thereby have

received an autographic letter from a man which I many years have honoured as one of the chief notabilities " in rebus magneticis." I preserve with delight, and perhaps a little vanity, letters from different English scientifical notabilities, as Sir Joseph Banks, Sir David Brewster, Professor Airy, Professor Forbes, General Sabine, Professor Barlow, and others; and to this treasure I now can add yours.

.   .   .   .   .   .

' Professor Oersted was a man of genius, but he was a very unhappy experimenter; he could not manipulate instruments.   He must always have an assistant, or one of his auditors who had easy hands, to arrange the experiment; I have often in this way assisted him as his auditor.   Already in the former century there was a general thought that there was a great conformity, and perhaps identity, between the electrical and magnetical force; it was only the question how to demonstrate it by experiments.   Oersted tried to place the wire of his galvanic battery perpendicular (at right angles) over the magnetic needle, but remarked no sensible motion.   Once, after the end of his lecture, as as he had used a strong galvanic battery to other experiments, he said, " Let us now once, as the battery is in activity, try to place the wire parallel with the needle;" as this was made, he was quite struck with perplexity by seeing the needle making a great oscillation (almost at right angles with the magnetic meridian). Then he said, " Let us now invert the direction of the current," and the needle deviated in the contrary direction.   Thus the great detection was made; and it has been said, not without reason, that " he tumbled

over it by accident." He had not before any more
idea than any other person that the force should be
transversal. But as Lagrange has said of Newton in
a similar occasion, " such accidents only meet persons
who deserve them.".

' You completed the detection by inverting the experiment by demonstrating that an *electrical current* can
be excited by a *magnet,* and this was no accident, but
a consequence of a clear idea. I pretermit your many
later important detections, which will conserve your
name with golden letters in the history of magnetism.

' Gauss was the first who applied your detection to
give telegraphic signals from the Observatory in Göttingen to the Physical Hall, in a distance of almost an
English mile from the Observatory.

' I very well understand your situation. I can also
not work in company with other persons; and I read
not much, for not to be distracted from my own way
of thinking; I allow that thereby many things escape
me, but I fear to be distracted upon sideways—" Non
omnia possumus omnes." Every one must follow his
own nature.

' I have translated an extract of your letter, and sent
it to Göttingen to Mr. Arndtsen.

' In the summer 1819, I visited, in long time, almost
every day the library in Royal Institution, in order to
extract magnetical observations (declination and in-
clination), from old works which our University was
not in possession of; for instance, " Hackluyt " and
" Purchas, his pillegrims," &c. So I am acquainted with
the place of your activity.

' I have in this year received your portrait from Mr.

Lenoir in Vienna, as also of Sir David Brewster. They shall decorate my study on the side of Oersted, Bessel, Gauss, and Struve.

'Believe me, Sir, sincerely your very respectful,
'CHR. HANSTEEN.'

FARADAY TO MISS REID.[1]

'Royal Institution: January 1, 1857. 5 o'clock P.M.

'My very dear Girl,—Your aunt has just brought me your letter; she has just had it. *We both* write by my pen, to save the post. Poor girl! we pity you all, as you may think; it needs not to say how much. The suddenness and awful character of the case may make us all tremble in our love to each other, and that whilst, as I trust it will be with you, we look unto Him who rules all things according to the purpose of His own will, let us strive to accept the sorrow submissively, and at the same time to do what remains in our power with hope of a blessing on the intention.

'How vain is life! In the midst of yours, which was not altogether smooth, still how great a trouble may be brought into it. But be composed; as far as remembrance of the hand that is over all can give composure, though it must be but partial. The Lord gave, the Lord taketh away. *Blessed be the name of the Lord.*

'As for advice, I can have none to give—those only who are on the spot can tell rightly; but we feel sorry it should come in ——'s way, though perhaps it may do no harm. —— we should have confidence in, except that her physical strength is weak, "but out of weakness are made strong" is a comfortable thought.

---

[1] On the sudden death of her sister.

'Give our kindest love to your father. In these
heavy sorrows, I think the words of condolence shrink
into my pen ; the thoughts of your heart must speak
for us ; and we commend you in your great trouble to
Him who is able to sustain you.

1858.
Æt.66–67.

' Your loving uncle,
' M. FARADAY.'

FARADAY'S HOUSE, HAMPTON COURT.

In 1858, through the thoughtful kindness of Prince
Albert, the Queen offered him a house on Hampton
Court Green.   It required repair, and he doubted
whether he could afford to do it up.

Faraday wrote to Dr. Becker, Prince Albert's secretary :—

' Albemarle Street, w.: April 20, 1858.

' My dear Dr. Becker,—I believe you know all about
the extreme kindness shown to me in respect of one of

Her Majesty's houses at Hampton Court.　I am in a
little difficulty about either accepting or declining it.
The manner in which it is offered to me is such as
would make it grievous to me to decline it, and yet, if
it is not improper, I should like to have a few words
with you before I finally settle.　At what hour could I
call to see you, and where?

'Ever your obliged,
'M. FARADAY.'

FARADAY TO DR. BECKER.

'Royal Institution: May 5, 1858.

'My dear Dr. Becker,—I had a most kind letter from
Col. Phipps, who, speaking in the name of Her Majesty,
removed all my difficulty, and so yesterday I could
accept the favour offered me.　I dare say you know
these things, but I felt as if I must either call or write
to you on the occasion; and as I had troubled you too
much already by calls, I take the latter course.　I am
surprised by the kindness I have received on this occasion, which, in the case of Her Majesty's unsought condescension, astonishes me.　I know that your good
wishes are with me in this matter, and they are of the
greater value to me, as they are free and unsolicited—
the spontaneous result of your own kind thought.
Whilst enjoying Hampton for a year or two, as I hope
to do, pleasant remembrances will be called up on
every side.

'Ever, my dear Dr. Becker, yours most truly,
'M. FARADAY.'

He wrote to a niece:—

'The case is settled. The Queen has desired me to dismiss all thoughts of the repairs, as the house is to be put into thorough repair both inside and out. The letter from Sir C. Phipps is most kind.'

To Sir C. Phipps he wrote :—

'I find it difficult to write my thanks or express my sense of the gratitude I owe to Her Majesty ; first, for the extreme kindness which is offered to me in the use of the house at Hampton Court, but far more for that condescension and consideration which, in respect of personal rest and health, was the moving cause of the offer. I feared that I might not be able properly to accept Her Majesty's most gracious favour. I would not bring myself to decline so honourable an offer, and yet I was constrained carefully to consider whether its acceptance was consistent with my own particular and peculiar circumstances. The enlargement of Her Majesty's favour has removed all difficulty. I accept with deep gratitude, and I hope that you will help me to express fitly to Her Majesty my thanks and feelings on this occasion.'

M. de la Rive was requested to write a short biography of Mrs. Marcet, and for this he asked if it were true that Faraday was inspired with his first taste for chemistry and physics by reading Mrs. Marcet's 'Conversations on Chemistry,' and whether this determined the course of his work. He answered :—

'Your subject interested me deeply every way, for Mrs. Marcet was a good friend to me, as she must have been to many of the human race. I entered the shop of a bookseller and bookbinder at the age of 13, in the year 1804, remained there eight years, and during the chief part of the time bound books. Now it was in those books, in the hours after work, that I found the beginning of my philosophy. There were two that especially helped me, the " Encyclopædia Britannica," from which I gained my first notions of electricity, and Mrs. Marcet's "Conversations on Chemistry," which gave me my foundation in that science.

'Do not suppose that I was a very deep thinker, or was marked as a precocious person. I was a very lively, imaginative person, and could believe in the "Arabian Nights" as easily as in the "Encyclopædia;" but facts were important to me, and saved me. I could trust a fact, and always cross-examined an assertion. So when I questioned Mrs. Marcet's book by such little experiments as I could find means to perform, and found it true to the facts as I could understand them, I felt that I had got hold of an anchor in chemical knowledge, and clung fast to it. Thence my deep veneration for Mrs. Marcet: first, as one who had conferred great personal good and pleasure on me, and then as one able to convey the truth and principle of those boundless fields of knowledge which concern natural things, to the young, untaught, and inquiring mind.

'You may imagine my delight when I came to know Mrs. Marcet personally; how often I cast my thoughts backwards, delighting to connect the past and the present; how often, when sending a paper to her as a thank-offering, I thought of my first instructress, and such like thoughts will remain with me.

'I have some such thoughts even as regards your own father, who was, I may say, the first who personally, at

Geneva, and afterwards by correspondence, encouraged, and by that sustained me.'

The work done in the laboratory in 1858 was very little. It was chiefly on the same subject as the previous year—time in relation to magnetism; but no 'successful result' was obtained.

To Tyndall's paper on the physical properties of ice, which was published in the 'Philosophical Transactions' this year, a note on ice of irregular fusibility was added by Faraday.

At the house of Mr. Gassiot, at Clapham, and also at the Institution, he worked on the stratification of the electric discharge in vacuum tubes. Mr. Gassiot says in his paper printed in the 'Philosophical Transactions' that he was much indebted to Dr. Faraday for suggestions, advice, and personal assistance in the progress of his researches.

For the Institution he gave two Friday discourses. The first was on static induction; and the other on Wheatstone's electric telegraph in relation to science (being an argument in favour of the full recognition of science as a branch of education). His notes of this lecture begin thus:—'I have no intention to look at electric telegraphs generally, and to decide amongst them, or to assign to those I shall consider their proper place. I take them as illustrations of the progress of scientific knowledge in our day; since the first dawn and the perfection of them is ours. By distinct steps I shall try to show what the world has gained in one branch only of scientific knowledge within a few short years; how that knowledge has stepped on by the joint application of intellect and experiment to

its development, and how quickly the practical application, covering with its abundant fruit the face of the
globe, has subserved the purposes of man in their most
intellectual as well as most practical exercises. I am
no poet, but if you think for yourselves, as I proceed,
the facts will form a poem in your minds.' His notes
end thus : ' We learn by such results as these, what is
the kind of education that science offers to man. It
teaches us to be neglectful of nothing, not to despise the
small beginnings—they precede of necessity *all great
things.* Vesicles make clouds; they are trifles light as
air, but then they make drops, and drops make showers
rain makes torrents and rivers, and these can alter the
face of a country, and even keep the ocean to its proper
fulness and use. It teaches a continual comparison
of the *small and great*, and that under differences
almost approaching the infinite, for the small as
often contains the great in principle, as the great does
the small; and thus the mind becomes comprehensive.
It teaches to deduce principles carefully, to hold them
firmly, or to suspend the judgment, to discover and obey
law, and by it to be bold in applying to the greatest
what we know of the smallest. It teaches us first
by tutors and books, to learn that which is already
known to others, and then by the light and methods
which belong to science to learn for ourselves and for
others ; so making a fruitful return to man in the future
for that which we have obtained from the men of the
past. Bacon in his instruction tells us that the scientific
student ought not to be as the ant, who gathers
merely, nor as the spider who spins from her own
bowels, but rather as the bee who both gathers and
produces.

All this is true of the teaching afforded by any part of physical science. Electricity is often called wonderful, beautiful; but it is so only in common with the other forces of nature. The beauty of electricity or of any other force is not that the power is mysterious, and unexpected, touching every sense at unawares in turn, but that it is under *law*, and that the taught intellect can even now govern it largely. The human mind is placed above, and not beneath it, and it is in such a point of view that the mental education afforded by science is rendered super-eminent in dignity, in practical application and utility; for by enabling the mind to apply the natural power through law, it conveys the gifts of God to man.'

He again gave the Juvenile Lectures on the metallic properties.

He made twelve reports to the Trinity House. The most important was on the electric light at the South Foreland. He went there with a Committee of the Trinity House, to see it from sea and land. The light was in the centre of the Fresnel apparatus, in the upper light, as a fixed light, and so comparable with the lower fixed light, which consisted of oil-lamps in reflectors. They went to the Varne light-ship. The upper was generally inferior to the lower light. Next morning they went to the lighthouse, and examined it by day; and again at night.

He was made Corresponding Member of the Hungarian Academy of Sciences, Pesth; and Sir David Brewster tried to tempt him to accept the Professorship of Chemistry in the University of Edinburgh.

The correspondence this year is not of great interest. His kind feeling is apparent, whether he writes to the

Duke of Northumberland, who wished to resign the Presidentship of the Royal Institution; or to a Royal Academician who wished to paint his portrait; or to Professor Tyndall who had just sent him an account of the ascent of Monte Rosa; or to his friend Schönbein, who he thought would help him to a subject for a Friday evening discourse at the Royal Institution.

FARADAY TO HIS GRACE THE DUKE OF NORTHUMBERLAND.

'Royal Institution: February 10, 1858.

' My Lord Duke,—According to your Grace's kind wish, Mr. Barlow has shown me the letter he received from you, and therefore I take the liberty of expressing an opinion upon *one point*, being sure, from the kindness your Grace showed me, when on a former occasion the matter of the Presidentship was in consideration, that I shall not by doing so give offence.

' I think it would be a very serious thing for the president and the secretary to resign at the same time. It would be sure to give occasion to the thought that there was some reason touching the character of the Institution, which united the two in the act; and the thought would be the more inconvenient because, as no open reason would appear, an unpleasant one, according to the common course of human nature, would be assigned. It grieves me to think that either president or secretary should ever leave the Institution; but as such events must occur in the course of nature, I do hope most earnestly that the resignation of the one may overpass the other by a year or two, that the present policy, which seems so good and prosperous, may not be suddenly interrupted, but transmitted through the gradual change.

Again begging for your Grace's kind reception of my free thoughts on this occasion, as on others, I sign myself, most truly, your Grace's free, faithful, and humble servant,

'M. FARADAY.'

1858.

Æt. 66.

FARADAY TO A R.A. WHO WISHED TO PAINT HIS PORTRAIT.

'Royal Institution : May 10, 1858.

' My dear Friend,—I am puzzled how to answer your note of the 8th, and our meeting in the evening only adds to my difficulty. The much occupation I have here, the continual delay of the pursuit of my own research (a delay now extending over two years), and the weariness of head and health resulting from this continued occupation, had made me resolve never to sit again.

' There are two or three who claim the first right if I break through the determination, but they are content with such an answer. Nevertheless, if I sit to you now, having done so very frequently, on two former separate occasions, I think they will have a just right to complain. I think you said that I had promised this to you. I do not remember any distinct promise, but as you have understood it so, I will endeavour to arrange for six sittings ; I trust that they will be sufficient to complete the last picture (which was left off very suddenly), and then, if I have health and strength, I must go to my research. I think, as far as I can see, that Tuesday mornings, early, would suit my arrangements.

' I understood you to say on Saturday, that it was for my sake you desired to paint another picture. Notwithstanding the high compliment which this

implies, all the reasons having relation to me are against
it. I should give offence to others, whom I esteem
most highly. I want my time, if well enough, for thought
and research ; exhausting as these are to me, I want
time for rest and health. Twice before have I for long
periods together been a burden to your genius ; and I
have arrived at such a time of life as to be no longer
vain of what may well be considered as a distinction.
I am quite prepared to do what I have said for *your*
sake, but find no motive in any circumstance that is
connected with *my own.*

'Ever yours faithfully,

'M. FARADAY.'

FARADAY TO DR. TYNDALL

(ON HIS ASCENT OF MONTE ROSA).

'Royal Institution: September 2,1858.

'My dear Tyndall,—I might not have written to
you again, but for the receipt of your letter by my
wife, detailing the ascent of Monte Rosa, and the enormous indiscretion I have committed thereupon. What
shall I say? I have sent it to the "Times." There, the
whole is out. I do not know whether to wish it may
appear to-morrow, or next day, or not. If you should
dislike it, I shall ever regret the liberty I have taken.
But it was so interesting in every point of view, showing the life and spirit of a philosopher engaged in his
cause ; showing not merely the results of the man's
exertions but his motives and his nature—the philosophy of his calling and vocation as well as the philosophy of his subject, that I could not resist ; and I was
the more encouraged to do so because, from the whole
character and appearance of the letter, it showed that

it was an unpremeditated relation, and that you had
nothing to do with its appearance, i.e. it will show that,
if it should appear. How I hope you forgive me.
Nobody will find fault with me, but you. It came too
late for the " Philosophical Magazine," but if the
" Times " does not put it in, I shall send it to the
" Philosophical Magazine." However, as this is only
the 3rd of the month, there is time enough for that.
  ' I won't give you any scolding. I dare say my
wife will when you see her. " Êtes-vous marié?"
indeed. I cannot help but feel glad you have done
it now it is done; but I would not have taken the
least portion of responsibility in advising you to such
a thing.
  ' I have no philosophy, and no news for you. I feel
just out of the world—forgetful, and dull-headed in
respect of science, and of many other things, but well
and content, as I have great reason to be.

  .   .   .   .   .

  ' Good-bye, my dear friend, ever truly yours,
                              ' M. FARADAY.

  ' Friday morning, 3rd. The letter is there.'

  To Professor Schönbein he thus writes, with respect
to a lecture to be given at the Royal Institution :—

FARADAY TO PROFESSOR SCHÖNBEIN.

' Royal Institution : November 13, 1858.

  'My dear Schönbein,—Daily and hourly am I thinking about you and yours, and yet with as unsatisfactory
a result as it is possible for me to have. I think
about ozone, about antozone, about the experiments

you showed to Dr. Bence Jones, about your peroxide of barium, your antozonised oil of turpentine, and it all ends in a giddiness and confusion of the points that ought to be remembered. I want to tell our audience what your last results are upon this most beautiful investigation, and yet am terrified at the thoughts of trying to do so, from the difficulty of remembering from the reading of one letter to that of another, what the facts in the former were. I have never before felt so seriously the evil of loss of memory and of clearness in the head; and though I expect to fail some day at the lecture table, as I get older, I should not like to fail in ozone, or in anything about you. I have been making some of the experiments Dr. Bence Jones told me of, and succeeded in some, but do not succeed in all. Neither do I know the *shape* in which you make them, as (I understand) good class experiments and telling proofs of an argument. Yet without experiments I am nothing.

'If I were at your elbow for an hour or two, I would get all that instruction (as to precaution) out of you, which might bring my courage up. I remember in old time (at the beginning of ozone), you charged me with principles and experiments. I wonder whether you could help me again? Most likely not, and it is a shame that I should require it; but without such help and precautions on my part, I am physically unable to hold my place at the table. And without I justify my appearance on a Friday evening, I had better withdraw from the duty.

'What I should want would be from *ten* to *fifteen*, or at most twenty, table experiments, with such instructions as to vessels, quantities, states of· solution, materials, and precautions, as would make the experiments visible to all, and certain and ready. Also the points of the general subject, in what you have found to be the *best order* for the argument and its proof.

'I have sought for the old bottle of antozone oil of turpentine, but believe I have used it all up. I fear it is of no use trying to make it by the end of January, next year. Yet about that time I must give the evening, if I give it at all. If you encourage me to give the argument (and I can only try if you help me), have you any of the substance you could spare, and could you find conveyance for it by rail or otherwise? I fear there is no other substance that will represent it, i.e., that approaches so near to isolated antozone, as that body does.

'Now do not scold me. I am obliged to speak as I do. Perhaps you had better tell me that I must *give up the subject*, for that I can hardly succeed in telling it properly by the way I propose. Do not hesitate to say so, for I am well prepared, by my inner experience in other matters, to suppose that may be the case. But then tell me so at once, that I may think on my position here for January.

'Now for a more cheery subject. . . . . . .

'Believe me to be, as ever, my dear Schönbein, your true and obliged friend,

'M. Faraday.'

FARADAY TO PROFESSOR SCHÖNBEIN.

'Royal Institution : November 25, 1858.

'Warmest thanks, my dear friend, for your last kind letter ; it has given me courage, and yet when I look into the journals about ozone, and see how many

things there are which have been said by different
men, and how thoroughly I have forgotten most of
them, it makes me very doubtful of myself, for I cannot hold many points in hand at once, as I used to do,
but I shall trust in your strength and kindness.

'It is the experimental proofs, and the method of
making them perfectly, about which I am anxious,
and none but the discovering philosopher himself
knows how best to make their value evident; for that
reason I desire to work with your tools, and in your
way.

'Kindest remembrances to the household, from one
always under obligation to you, and ever yours,

'M. FARADAY.'

In 1859 the notes made in the laboratory book
are characteristic in the highest degree of the working
of Faraday's mind.   In them may be seen the depths
of his imagination, brought to the test of the most
searching experiments, the strongest faith in his
thoughts with the truest judgment of his results.   On
February 10, paragraph 15,785, he writes:—'Surely
the force of gravitation, and its probable relation to
other forms of force, may be attacked by experiment.
Let us try to think of some possibilities.

'15,786. Suppose a relation to exist between gravitation and electricity, and that as gravitation diminishes
or increases by variation of distance, electricity either
positive or negative were to appear—is not likely;
nevertheless, try, for less likely things apparently have
happened in nature.

'15,787. There is more chance of any observable
effect in a body acted on by the earth, than in the

same body acted on by a like body. There is more
chance of a variation being observed in a ton of water
or lead, when lifted a hundred yards upwards from
the earth, than in the same ton when moved a hundred yards in a horizontal direction from the side of
another ton, by which it at first stood.

'15,788. Must not be deterred by the old experiments (10,018, &c.) If there be any true effect of
gravity, it may take much gravitating matter to make
the effect sensible, and I had but very little. Moreover, the action of a body with or against gravity
ought not to form a current in a closed circuit, as tried
in the former case, but perhaps give opposite states
in lifted or depressed bodies; and though a current
might be found in a wire connecting two such, it
would not be a current in a circuit. So may consider the imaginable effects under two views, *static*
and *dynamic*. Take the former first, and imagine as
follows :—

'15,789. If an insulated body, being lifted from the
earth, does evolve electricity in proportion to its loss
of gravitating force, then it may become charged to a
very minute degree, either *positive* or negative. When
thus charged, it may be discharged, and then, if
allowed to descend insulated, it would become charged
in the opposite matter, and so on. If three or more
bodies of the same size, but in weights, as 1, 2, and 3,
then the intensity of the charge ought to be as the
densities.

'15,790. Might not two globes (or masses, as pigs of
lead), A, B, attached to the end of a long rope, passing
over a large pulley at the top of the clock-tower, or
in the whispering-gallery of St. Paul's, serve an ex-

perimental purpose? Starting with both balls insulated, discharged, and balanced, then it would be
easy to raise B and lower A, and examination by a
very delicate static electrometer might show A charged
positive, and B negative; then discharging both, and
reversing the motion, B would come down positive,
and A become negative, and so on. The static electrometer might be applied either above or below, or at
both places. If the effect were real but insensible,
several journeys up and down might be effected, the
discharge above being made by bell-wire and touching
lever, or the discharge above and below might be
made *automatically* to two electroscopes, one above
and one below, so as to accumulate many results into
one. These electrometers being very delicate and of
the condensing kind, one man, having only to turn a
windlass, might work the apparatus for half a day, or
it might be kept in motion by a steam-engine, or other
mechanical power.

'15,795. The evolution of *one* electricity would be a
new and very remarkable thing. The idea throws a
doubt on the whole; but still try, for who knows what
is possible in dealing with gravity.

'15,796. The first thought would give a new relation, a relation of a dual power to a single power,
which would probably give a modification to the
character of singleness supposed to belong to gravitation, for it would then be as dual as electricity.

'15,799. Perhaps a jet of drops of water from a height
would tell below, only water is a bad substance, because
of the discharging facility of moist air.

'15,800. Possibly a jet of lead would do better; the
fall of shot in the shot-tower; might insulate the tub of

water into which it falls below, and so get traces of any evolution.

' 15,804. Let us encourage ourselves by a little more imagination prior to experiment. Atmospheric pheno-mena favour the idea of the convertibility of gravitating force into electricity, and back again; probably (or perhaps then into heat) matter is constantly falling and rising in the air. The difference and the change of place of the bodies subject to gravity would perhaps give a predominant electric state above, as the negative, but also an occasional *charge* of the other state, the positive. If there be this supposed relation of gravity and electricity, and the above space be chiefly or generally negative, then we might expect that, as matter rises from the earth, or moves against gravity, it becomes negative.

' 15,805. Here we might expect a wonderful opening out of the electrical phenomena.

' 15,806. So to say, even the changed force of gravity as electricity might travel about the earth's surface, changing its place, and then becoming the equivalent of gravity.

' 15,807. Perhaps heat is the related condition of the force when change in gravity occurs; atmospheric phenomena are not at first sight opposed to this view; might associate a thermo-electric pile or couple, to see if change of elevation from the earth causes any sensible change of temperature.

' 15,808. Perhaps almost all the varying phenomena of atmospheric heat, electricity, &c., may be referable to effects of gravitation, and in that respect the latter may prove to be one of the most changeable powers, instead of one of the most unchanged.

' 15,809. Let the imagination go, guarding it by judgment and principle, but holding it in and directing it by *experiment*.

' 15,810. If any effect, either electric or calorific, then consider the infinity of action in nature—a *planet* or a *comet* when nearer or further from the sun. Dr. Winslow's observations on earthquakes, a falling river or cascade, the Falls of Niagara, evaporation, vapour rising, rain falling, hail, negative state of the upper regions, condition of the inner and deeper parts of the earth, their heat, a falling star or aerolite heated, a volcano and the volcanic lightning, smoke in a chimney perhaps goes out electrified.

' 15,811. What a multitude of events and changes in the atmosphere would be elucidated by such actions. I think we have been dull or blind not to have suspected some such results.

' 15,814. If anything results then we shall have

' 15,815. An entirely new mode of the excitement of either heat or electricity.

' 15,816. An entirely new relation of natural forces.

' 15,817. An analysis of gravitation force.

' 15,818. A justification of the conservation of force.'

On March 4, he gets his thoughts more closely to the experiments he intends to make, and on the 10th he begins his experiments, and continues them on the 11th; on the 12th he was at the clock-tower of the Houses of Parliament. ' It will be excellent for my purpose,' he says. He continued his experiments at the staircase of the Royal Institution on the 14th, 15th, 16th, 17th, 18th, 19th, 21st, 22nd, 23rd and 24th. On the 26th he went to the shot-tower and found that the men do not work on Saturday afternoon; so on April 11 he goes for further data for making experiments. He makes this note :—

'15,915. It would be strange if a body should heat as gravitation increases by nearness of distance. We conceive of heat as a positive force, and of gravitation as a positive force, and then, instead of being the inverse of each other, they would seem to grow up together; or else heat must be negative to gravity, or the converse of gravity, and gravity must be in the same negative or converse relation to heat. This is against the expectation of anything from the heat experiment. Nevertheless, make it, for, who knows, if gravitation depend upon forces external to the particles, such results might happen. Try.' On the 12th, 15th, 16th, 17th, and 18th of April, May 3rd, 5th, 9th, 16th, 17th, 19th, 20th, 23rd, 26th, 27th, 28th, 30th, 31st, June 4th, 6th, and 11th, he continues at work.

He says, paragraph '15,985. So there is no evidence, by either apparatus, that any difference due to gravity, varying by an elevation of 165 feet, can show a relation to heat by causing a change of temperature.'

On July 9, he was 'at the shot-tower, to try for electricity.'

A pig of lead was charged, then sent to the top of the tower, 165 feet, then lowered and examined; it still gave a fair charge of electricity, though the charge had in some degree diminished. Another hundredweight of lead was added, 'still there were no clear signs of any effect.'

'15,997. The experiments were well made, but the results are negative.'

These results were made into a 'Note on the Possible Relation of Gravity with Electricity or Heat,' dated

April 16, 1860. This was received by the Royal Society on June 7, but the opinion of Professor Stokes was 'against sending it in for the "Transactions,"' because it contained only negative results. It was consequently withdrawn, and never was published.

It begins thus : ' Under the full conviction that the force of gravity is related to the other forms of natural power, and is a fit subject for experiment, I endeavoured on a former occasion ("Phil. Trans.," 1851, p. 1), to discover its relation with electricity, but unsuccessfully. Under the same deep conviction, I have recently striven to procure evidence of its connection with either electricity or heat. By a relation of forces I do not mean the production of effects associated usually with one form of power, by the exercise of another, unless the results are direct. There is no difficulty in obtaining either electrical or heating phenomena, indirectly from the force of gravity; but such relations as those between electricity and magnetism, or electricity and heat, or chemical force and electricity, have not yet been obtained as respects gravity and other forces ; and though I have again failed, I do not think either the trials or the views which led to them are without interest. The former are not more than negative, and as such give no proof that the latter are inconsistent, or in a wrong direction. It might, indeed, have been anticipated, from the very views I entertain, that we could hardly hope to lay hold by experiment of such an amount of gravitating force as would yield appreciable evidence of electric or heat force ; but if we were to stop the first institution of experiment in any new direction, for such a reason, what progress—or, more, what discovery—could we hope to make by its means in any of the as yet un-discovered paths of science?

. . . . . .

'I proceeded to make experiments by raising and lowering masses of matter, and testing their states at the upper and lower stations.'

The note ends : ' Though these results are negative, and though the truth of nature may be that there is no such relation as that I have been looking for, yet I cannot accept them as conclusive ; and if the opportunity should arise of making the experiments with such electrometers as Dellmann's or Thomson's, I should ask Mr. Walker again to let me repeat the electrical experiments in his (shot) tower.'

The most important work that was done in 1859 was for the Trinity House.   Eleven reports were made, and one was sent to the Board of Trade.   The chief subject was on the use of the electric light at the South Foreland.

In 1857, Faraday said at the end of a report on Professor Holmes's magneto-electric light : ' I hope a situation may be selected where the magneto-electric lamp can be safely and effectually tried for a time, and under circumstances during which all the liabilities may be thoroughly eliminated.   The light is so intense, so abundant, so concentrated and focal, so free from under-shadow (caused in the common lamp by the burner), so free from flickering, that one cannot but desire it should succeed.   But it would require very careful and progressive introduction, men with pecu-liar knowledge and skill to attend it, and the means of instantly substituting one lamp for another in case of accident.   The common lamp is so simple both in principle and practice, that its liability to failure is very small.   There is no doubt that the magneto-electric lamp involves a great many circumstances tending to make its application more refined and delicate, but I would fain hope that none of them will prove a barrier to its introduction.   Nevertheless, it must pass into practice only through the ordeal of a full, searching, and prolonged trial.'

On December 8, 1858, the light was exhibited at the Upper South Foreland lighthouse.   Faraday examined it on shore, and on the sea, and reported to the Trinity House.   In consequence, Professor Holmes rearranged his apparatus, and on March 28, 1859, the exhibition was resumed.   On April 20 and 21 it was again examined by him from sea and land.   ' At sea, the upper light was always superior to the lower, of a white colour, and very steady, and not changing except near the astragals and bars.   It has often been described as a flickering light, but it was not so to-night.   As we went in towards the lights the electric light rose up beautifully ; and certainly as a light is unexceptionable, and superior much to the central lamp.   The increase of light was very steady, and as we went up towards the lighthouse, i.e. to Dover, the effect was in every respect highly satisfactory.'   The following night he examined the light on shore.   ' The light from within was beautiful, but seemed wonderfully small.   It is a perfect electric light.'   ' The light was beautifully steady and bright, with no signs of variation.   The appearance was such as to give confidence to the mind.   No doubt about its continuance.   As a light, it is unexceptionable ; as a magneto-electric light, wonderful ; and seems to have all the adjustments of quality, and more than can be applied to a voltaic electric light or a Ruhmkorff.'

All the reports sent in from the surrounding light-houses, floating lights, and pilot vessels, confirmed the superiority of the upper light above the lower; and on April 29, Faraday reported ' that Professor Holmes has practically established the fitness and sufficiency of the magneto-electric light for lighthouse purposes, so far as its nature and management are concerned. The light produced is powerful beyond any other that I have yet seen so applied, and in principle may be accumulated to any degree. Its regularity in the lantern is great, its management easy, and its care there may be confided to attentive keepers of the ordinary degree of intellect and knowledge.

'There are many other considerations beyond this establishment of the fitness of the light in principle and management for lighthouse purposes, regarding its introduction into lighthouses generally, on which I should hesitate to speak before those who are far more competent to judge of these matters than I am, were it not for the encouragement which the brethren of the Trinity House give me, and especially as regards this light.'

And then he points out some facts which are against, and others in favour of this light.

' Against it is its comparative complexity, and its outfit and current expenditure. In favour of it is the increase of light, and the advancement of lighthouses in power.' In conclusion, he says : ' I must bear my testimony to the perfect openness, and candour, and honour of Professor Holmes. He has answered every question, concealed no weak point, explained every applied principle, given every reason for a change either in this or that direction, during several periods of close questioning, in a manner that was very agreeable to me, whose duty it was to search for real faults and possible objections, in respect both of the present time and the future.'

Early in the following year, when the trial time was nearly ended, Faraday again went to Dover, but the snow was so deep in drifts and cuttings, that he could not get to the lighthouse. He tried again the next week, and with difficulty he succeeded.

On the 28th he wrote to the Trinity House, after examining the reports :—

FARADAY TO THE SECRETARY.

' Sir,—The appointed time during which the magneto-electric light was to be placed under practical trial at the South Foreland having come to an end, I may be allowed to say that it has so far justified itself in its results as to make me hope that the Trinity House will see fit to authorise its application, either there or somewhere else, for a further and a longer period. The light has proved to be practicable and manageable, and has supplied the means of putting into a lighthouse lantern, for six months or more, a source of illumination far surpassing in intensity and effect any other previously so employed. I do not know at what cost this result has been obtained, but unless this is very great indeed, I am of opinion that a large increase upon the expense of the old method (estimated upon the quantity of light obtained), ought to be permitted in the first

establishment of a mode of illumination which appa- 1859.
rently promises many improvements during its farther ÆT.67–68.
development.

'M. FARADAY.'

For the Trinity House also he reported this year on
Way's mercurial electric light; the one advantage
it had was that the place of the light was unchangeable.

For the Institution, after Easter, he gave a course of
six lectures on the various forces of matter, and he
took two Friday evening discourses. For the third
time, he chose Schönbein's ozone and antozone as one
subject; his other discourse was on phosphorescence
and fluorescence. He begins his notes of this last lecture
thus: 'Nothing is more fitted to enlarge the views of
a philosopher, and liberate him from preconceptions,
than the study of what may appear to be the anomalous
habitudes of a single agent; and no agent is more
striking in that respect than light. New researches,
therefore, are good.'

At Christmas he gave the Juvenile Lectures on the
forces of matter.

This year he was asked, by Mr. W. Smith, to publish
the course of Juvenile Lectures on the metallic properties.

He replied :—

'Royal Institution : January 3, 1859.

'Dear Sir,—Many thanks to both you and Mr. Bentley. Mr. Murray made me an unlimited offer like that
of Mr. Bentley's many years ago, but for the reasons I
am about to give you I had to refuse his kindness.
He proposed to take them by short-hand, and so save

me trouble, but I knew that would be a thorough failure ; even if I cared to give time to the revision of the MS., still the lectures without the experiments and the vivacity of speaking would fall far behind those in the lecture-room as to effect. And then I do not desire to give time to them, for money is no temptation to me. In fact, I have always loved science more than money ; and because my occupation is almost entirely personal, I cannot afford to get rich.

'Again thanking you and Mr. Bentley, I remain, very truly yours,

'M. Faraday.'

He was one of a Commission appointed to consider the subject of lighting public galleries by gas ; and he reported favourably on the experimental attempt at the Sheepshanks Gallery.

Among the letters of this year one from Mrs. Somerville shows her faith in his powers of research ; one to the Rev. John Barlow is on the new degrees in science at the University of London ; and two letters to relations give his thoughts on religious questions—thoughts which, though often in his mind, were very rarely to be seen in his writing, or to be heard in his ordinary conversation.

MRS. SOMERVILLE TO FARADAY.

'Florence : February 1, 1859.

'My dear Dr. Faraday,—I cannot tell you how much I have been delighted and gratified by your letter, and by your kind acceptance of my book. I should not have dared to send it to you from any merit it may

have in itself, but I have no other way of offering the   1859.
tribute of my most sincere and heartfelt admiration of   Æt.67–68.
your transcendent discoveries of the laws and deep
mysteries of nature.

'I fear from what you say that I may have expressed
myself ambiguously with regard to your views of gravitation. I certainly did not mean to do so, for, on the contrary, they convey to my mind the most perfect conviction, and I only hope you may live to complete what
Newton began, by the discovery of that one comprehensive power of which gravity and all the correlative and
convertible forces are but parts. Meanwhile, I wish
you success in your research for time in magnetism,
which there can be no doubt you will accomplish,
having already so beautifully connected magnetism
with light, whose velocity is known.

I fear you tax your health too severely; subjects so
abstruse as you are accustomed to consider must fatigue
even your mind, which makes occasional repose necessary; so I wish you would come here and amuse yourself for a little; we should be indeed delighted to see
you, and there are many things that would interest you.

'Many thanks for the volume of your papers and
researches, which you intend to send to me; it will be
a very precious gift. Mr. Somerville and my daughters
desire to be kindly remembered to you, and be assured
that I am ever, with sincere friendship, yours,

'MARY SOMERVILLE.'

Professor John Phillips, Museum House, Oxford,
asked him to subscribe to a testimonial to Humboldt.

FARADAY TO PROFESSOR JOHN PHILLIPS.

'Royal Institution, London, w. : October 21, 1859.

'My dear Phillips,—I have received your letter of September 15. I have several times considered very carefully the matter it refers to, namely, the Humboldt testimonial, and I cannot bring myself to think that it is a step in the right direction. Humboldt's memory cannot, according to my view, be honoured by any act of the kind. A feeling of the highest and finest character belongs to the name, and in my opinion that feeling is lowered by the association of the name with anything partaking of the character of a testimonial. No such act can in any shape do honour to Humboldt's memory, for that stands alone in its glory. On the contrary, the system has been abused so frequently of late, that I for one feel Humboldt's name would be hurt by association in any way with it.

As to the aid that may come to science by the means proposed, I do not think that any who may be willing to yield it would not do so as freely for science's own sake as for the sake of an oblique and posthumous association with the name of Humboldt. Indeed, I cannot bring my mind to the conclusion that Humboldt himself would, if he were in the flesh, approve of such a motive and manifestation.

' Nevertheless, doubting my own judgment, and seeing how many appear to be in favour of the proceeding, of whose judgment and feeling I cannot but think most highly, I beg to fill up the paper you have sent me for the sum of 5l., and enclose it in this letter. I hope

you will bear with these remarks. I should not have
felt true to you and myself if I had not made them.

' Ever, my dear Phillips, most truly yours,
' M. FARADAY.'

To Dr. Tyndall, who asked whether he should refuse
the Professorship of Physics at Edinburgh, he writes :—

FARADAY TO DR. TYNDALL.

'Edinburgh : November 15, 1859.

' My dear Tyndall,—I really cannot advise you ; I
can only tell you what I should do, and what I did do
under like circumstances. When the Chair of Chemistry
was offered me, under the strongest private assurances
of the authorities, I declined it.   It was all a matter of
feeling with me, but the feeling was, that if I had a
sufficient moderate income in London, nothing would
make me change London for Edinburgh.   Others
might reverse the terms, and say nothing should bring
them to London, so that really I have no advice to give,
for I suppose I may now assume that you have a
competency in London, and that all beyond will
come under the points of honour, prosperity, and
pleasure.

' Ever, my dear Tyndall, truly yours,
' M. FARADAY.'

To the Rev. John Barlow he writes on August 10,
from Hampton Court :—

FARADAY TO REV. JOHN BARLOW

' As I have been out here with only runs into town,
I really know very little of what is going on there, and

what I learn I forget. The Senate of the University accepted and approved of the report of the Committee for Scientific Degrees;[1] so that that will go forward (if the Government approve), and will come into work next year. It seems to give much satisfaction to all who have seen it, though the subject is beset with difficulties ; for when the depth and breadth of science came to be considered, and an estimate was made of how much a man ought to know to obtain a right to a degree in it, the amount in words seemed to be so enormous as to make me hesitate in demanding it from the student; and though in the D.S. one could divide the matter and claim eminence in one branch of science, rather than good general knowledge in all, still in the B.S., which is a progressive degree, a more extended though a more superficial acquaintance seemed to be required.   In fact, the matter is so new, and there is so little that can serve as previous experience in the founding and arranging these degrees, that one must leave the whole endeavour to shape itself as the practice and experience accumulates.'

In the winter, Professor Schönbein's daughter died in London. Faraday refers to this in a letter to his friend.

FARADAY TO PROFESSOR SCHÖNBEIN.

'Royal Institution: April 25, 1859.

'My dear Schönbein,—I am glad you went out, for though all things would be distasteful to you, still they work out the transition back again from sudden and deep grief to a more collected, healthy, and neces-

---

[1] Faraday, Mr. Hopkins, and Dr. Carpenter drew up the first defined scheme for the examinations.

sary state of mind. For the same reason I am very
glad that Mrs. Schönbein has left home for a little while,
and trust that it may calm her spirits and do her good.
It is impossible for me to write to you, or do anything connected with you, without thoughts of your
dear daughter entering in.

.   .   .   .   .   .

' I gave a Friday evening on ozone and antozone, for
which only a few weeks before I had given tickets at
her request to some friends of hers, but I could not
and cannot talk to you about it. I did my best,
though with thoughts often pressing in ; still let me
thank you for what you had before the sad event done
to help me.

' My wife joins me in kindest remembrances and
thoughts ; and so, too, does my niece, for though she
was not much known to you, yet she was to Miss
Schönbein. Extend these sympathising thoughts to the
children who remain to comfort you.

' Ever, my dear Schönbein, yours,

' M. FARADAY.'

FARADAY TO HIS NIECE MRS. DEACON.

' The Green, Hampton Court, s.w : August 12, 1859.

' My dear C.,—I am a little tired, dull, and
unable to work, or even to read ; so I write to you. I
have your letter before me, and so that is a moving
cause ; and it is rather grave, and that renders the cause
more effectual. I never heard of the saying that separation is the brother of death ; I think that it does
death an injustice, at least in the mind of the Christian ;
separation simply implies no reunion ; death has to the
Christian everything hoped for, contained in the idea of

reunion. I cannot think that death has to the Christian anything in it that should make it a rare, or other than a constant, thought ; out of the view of death comes the view of the life beyond the grave, as out of the view of sin (that true and real view which the Holy Spirit alone can give to a man) comes the glorious hope ; without the conviction of sin there is no ground of hope to the Christian. As far as he is permitted for the trial of his faith to forget the conviction of sin, he forgets his hope, he forgets the need of Him who became sin, or a sin-offering, for His people, and overcame death by dying. And though death be repugnant to the flesh, yet where the Spirit is given, to die is gain. What a wonderful transition it is! for, as the apostle says, even whilst having the first-fruits of the Spirit, the people of God groan within themselves, " waiting for the adoption, to wit, the redemption of the body." Elsewhere he says, that whilst in " the earthly house of this tabernacle we groan earnestly, desiring to be clothed upon with our house which is from heaven."

' It is permitted to the Christian to think of death; he is even represented as praying that God would teach him to number his days. Words are given to him, " O grave, where is thy sting? O death, where is thy victory ? " and the answer is given him, " Thanks be to God, who giveth us the victory through our Lord Jesus Christ." And though the thought of death brings the thought of judgment, which is far above all the trouble that arises from the breaking of mere earthly ties, it also brings to the Christian the thought of Him who died, was judged, and who rose again for the justification of those who believe in Him. Though the fear of death be a great thought, the hope of eternal

life is a far greater.   Much more is the phrase
the apostle uses in such comparisons.   Though sin
hath reigned unto death, much more is the hope of
eternal life through Jesus Christ.   Though we may
well fear for ourselves and our faith, much more may
we trust in Him who is faithful; and though we have
the treasure in earthen vessels, and so are surrounded
by the infirmities of the flesh with all the accompanying hesitation — temptations and the attacks of the
adversary—yet it is that the excellency of the power
of God may be with us.

'What a long, grave wording I have given you; but
I do not think you will be angry with me.   It cannot
make you sad, the troubles are but for a moment; there
is a far more exceeding and eternal weight of glory for
them who, through God's power, look not at the things
which are seen, but at the things which are not seen.
For we are utterly insufficient for these things, but the
sufficiency is of God, and that makes it fit for His
people—His strength perfect in their weakness.

'You see I chat now and then with you as if my
thoughts were running openly before us on the paper,
and so it is.   My worldly faculties are slipping away
day by day.   Happy is it for all of us that the true good
lies not in them.   As they ebb, may they leave us as
little children trusting in the Father of mercies and
accepting His unspeakable gift.

'I must conclude, for I cannot otherwise get out of
this strain; but not without love to Constance, and
kindest remembrances to Mr. Deacon.

'Ever, your affectionate uncle,
'M. FARADAY.'

He was asked to write his opinion on the Revivals, &c.

'The Revivals, &c., cannot trouble the Christian who is taught of God (by His Word and the Holy Spirit) to trust in the promise of salvation through the work of Jesus Christ. He finds his guide in the Word of God, and commits the keeping of his soul into the hands of God. He looks for no assurance beyond what the Word can give him, and if his mind is troubled by the cares and fears which may assail him, he can go nowhere but in prayer to the throne of grace and to Scripture. No outward manifestation, as of a revival, &c., can give either *instruction* or *assurance* to him, nor can any outward opposition or trouble *diminish his confidence* in "Christ crucified, to the Jews a stumbling-block, and to the Greeks foolishness; but to them who are *called*, Christ the power of God, and the wisdom of God." If his attention is called to the *revivals*, it cannot be that he should feel instruction there or assurance there, other than what he finds in the Scriptures, without reference to them; and it seems to me that any power they may have over his mind other than the Scripture has, must be delusion and a snare.

'That man in his natural state is greatly influenced by his fellow-creatures and the forms of emotion which are amongst them, is doubtless true, even when it concerns what he considers his eternal welfare. How else would the wonderfully varied and superstitious forms of belief have obtained in the world? What carries the Mormons into the desert, surrounded by trouble and the enmity of those around them? What sustains a spiritual dominion like the Papacy, aided by the nations around it, to proclaim the name of Christ whilst it contradicts His

Word—refuses it (the record of the Spirit) to the people —and crushes out with all intolerance the simple obedience of the truth? Man's natural mind is a very unstable thing, and most credulous, and the imagination often rules it when reason is thought to be there. Mesmerism has great power over it; so has poetry; so has music; so has the united voice of the multitude; so have many other things: but these things are, so to say, indifferent as respects the *character* of the object they may be used to sustain, and are just as powerful in favour of a bad cause as a good one. Among the contradictory and gross systems of religion, or the numerous and opposed systems of political government, any one of them may be sustained by the use of such agencies as these.

' The Christian religion is a revelation. The natural man cannot know it. He, not knowing it, is liable in respect of religion to all the influences before mentioned, finds in them snares and delusions, and either becomes an infidel or is subject to every wind of doctrine. The Christian religion is a revelation, and that revelation is in the Word of God. According to the promise of God, that Word is sent into all the world. Every call and every promise is made freely to every man to whom that Word cometh. No revival and no temporal teaching comes between it and him. He who is taught of the Holy Spirit needs no crowd and no revival to teach him; if he stand alone he is fully taught, for the Comforter (the Spirit) taketh of the things of Christ and showeth them to His people. And if in the mercy of God it *should* please Him that one seeing the commotion about him should be led to examine his ways, it will only be in the Word of the testimony, the Word of God, that he will find the revelation of the new and living way by which he may rejoice in hope of entering the Kingdom of Christ.'

The work in 1860 was not important; the experiments in the laboratory were made chiefly in February, and were sent to the Royal Society, and published in the 'Proceedings' as a paper on regelation.

For the Institution he gave lectures on two Friday evenings; one on lighthouse illumination, the other on the electric silk-loom.

He said: 'There is no part of my life which gives me more delight than my connection with the Trinity House. The occupation of nations joined together to guide the mariner over the sea, to all a point of great interest, is infinitely more so to those who are concerned in the operations which they carry into effect; and it certainly has astonished me, since I have been connected with the Trinity House, to see how beautifully and how wonderfully shines forth among nations at large the desire to do good.

.   .   .   .   .   .

' I will not tell you that the problem of employing the magneto-electric spark for lighthouse illumination is quite solved yet, although I desire it should be established most earnestly, for I regard this magnetic spark as one of my own offspring. The thing is not yet decidedly accomplished, and what the considerations of expense and other matters may be, I cannot tell. I am only here to tell you as a philosopher how far the results have been carried, but I do hope that the authorities will find it a proper thing to carry out in full. If it cannot be introduced at all the lighthouses, if it can only be used at one, why, really it will be an honour to the nation which can originate such an improvement as this, one which must of necessity be followed by other nations.

'You may ask what is the use of this bright light.

.　　.　　.　　.　　.　　.

'This intense light has therefore that power, which we can take advantage of, of bearing a great deal of obstruction before it is entirely obscured by fogs or otherwise.'

He ended thus :—

'Taking care that we do not lead our authorities into error by the advice given, we hope that we shall soon be able to recommend the Trinity House, from what has passed, to establish either one or more good electric lights in this country.' The Trinity House decided to continue the trial of the electric light. Dungeness was chosen, although, from its higher position, Faraday recommended the Start Point.

At Christmas he gave his last course of Juvenile Lectures on the chemical history of a candle.

These Juvenile Lectures were begun at the Institution in 1825, and Faraday gave his first course in 1827 on chemistry. Altogether, he lectured at Christmas for the Institution nineteen years. The last ten years these lectures were given yearly by him without any interruption; usually they were on electricity or on chemistry. Twice he gave the chemical history of a candle, and three times he took as his subject the forces of matter.

With Sir Roderick Murchison, he was requested by Mr. W. Cowper to report upon the means of preserving

the stonework of the New Palace of Westminster. He
was 'to form some kind of judgment respecting the
preservation of the stonework.' 'I was to be guided by
the appearance and state of the prepared specimens, and
these alone.' Three processes were under consideration,
those of Mr. Szerelmey, Mr. Ransome, and Mr. Daines.
Mr. Daines called on him at the Royal Institution.
Faraday writes: 'Mistook him for another person, and
saw him for a few moments. Refused to discuss with
him. He threatened legal proceedings. Sent him out,
and all his papers.' Mr. Ransome began a written discussion with him, which Faraday ended by saying:
'I have no intention of altering my position in respect
of the First Commissioner of Works, or transferring my
correspondence from him to any other person.' Mr.
Cowper wanted him to analyse Mr. Szerelmey's preparation. He answered: 'It is especially proper that I
should not do so, for I was bound at first to give an
opinion *without* knowing the composition, and I would
rather not alter my opinion now. If you consider an
analysis necessary for your object, I conclude that some
of the professional men attached to the Government at
the Jermyn Street Museum, Woolwich, or elsewhere, will
be the proper persons to undertake it. For my own
part, I think time is the *only* test of such a practical
matter.

'I have lately had a visit and a threat of legal proceedings from Mr. Daines, on account of my answers
to your questions. I will candidly confess that such
results cool in some degree my willingness to answer
all inquiries made of me by the governmental boards.
If I thought that such a case were likely to occur again,
I would make *all* my letters *private*, to prevent like

results. Whenever you give me the pleasure of being any way useful to you again, I hope you will help me to keep clear of the parties, whose object is, of course, profit.'

Mr. Cowper was distressed to think how much trouble and annoyance had been caused to Faraday by his kindness in undertaking to judge of the indurating process, and ends saying he 'will not ask too much of him again.'

He was made Foreign Associate of the Academy of Sciences, Pesth, and Honorary Member of the Philosophical Society of Glasgow.

He resumed the office of elder in his church in the autumn, and after holding it for about the same time as he did before, three years and a half, he finally resigned it.

At Whitsuntide he was at Hampton Court. A storm spent some of its violence on the trees in Bushy Park, and the words in which he describes the experiments which nature made for him in the country, are a simple sermon on the vanity of life, showing how beauty and glory may suddenly pass away and serve only for the momentary amusement of the thoughtless crowd.

He made the following entry in his niece's journal :

'This Whit Monday will be long remembered among Hampton Court holiday folk. The wind from the west was very strong, so as to blow down and break trees, displace nests, tear off the mistletoe from the limes, and that whilst the sports were going on. Sunshine and rain alternated, but the rain was only in showers, though some of these were heavy.   In the morning, a tree upon the right at the entrance of Bushy Park was blown up by the roots, but clinging at the trunk on one side, it swerved in its fall, and coming against another tree, sheared off all its large branches on one side, and covered the ground with a sad ruin of bright green leaves, fresh horse-chestnut flowers and fragmented stems, the broken ends of which hung very white and fresh, and shone forth like pearls amongst the green and flowery setting.   When we saw it about 12 o'clock, multitudes of visitors were climbing into, over, and about it, some were gathering flowers, others a stick, others wandering in and out and under the branches, hiding themselves amongst the leaves.   Some of these people had heard and seen it fall.   In the afternoon, we went again to look at the tree, and then to see into what a state the people had brought it.   Its glory was low before ; but now to see how it was abased ! All the flowers gone ; such stem parts as had projected into the air had served for horses as long as they could bear the weight added to them, and now were broken down. In the morning, it was a beautiful ruin, and, with the people among it, would have made a fine picture ; now it was a degraded and almost formless heap.   And behold another tree on the left-hand side of the gate, as large as the former, has been blown down, happily without injury to any one.   The tree snapped off short about eight feet from the ground ; in its fall it also treated a neighbouring tree very sadly, and together they made a sad heap of ruins.   Besides these trees, great branches were lying about under other trees in all parts.   The corner tree on the right of our house on the green had a very large branch torn off ; so also had another tree on the green, towards the west and north-west part. Branches were torn off the trees in Mr. Roberts' ground and thrown into our garden, but happily the greenhouse escaped. I was in and out much. Margery also rambled; altogether, it was a day to be remembered.

'The numbers visiting the Palace are said to be unprecedentedly great.'

1860.

_Æt._ 68.

FARADAY TO PROFESSOR SCHÖNBEIN

(THE BURIAL PLACE OF WHOSE DAUGHTER, FARADAY HAD MARKED IN THE NORTH LONDON CEMETERY).

'Royal Institution: March 27, 1860.

'My dear Schönbein,—It seems to me a long time since we have spoken together, and I know that the blame is mine, but I cannot help it, only regret it, though I can certainly try to bring the fault to an end. When I want to write to you it seems as if only nonsense would come to mind, and yet it is not nonsense to think of past friendship and dear communions. When I try to write of science, it comes back to me in confusion ; I do not remember the order of things, or even the facts themselves. I do not remember what you last told me, though I think I sent it to the "Phil. Mag.," and had it printed; and if I try to remember up, it becomes too much, the head gets giddy, and the mental view only the more confused. I know you do not want me to labour in vain, but I do not like to seem forgetful of what you tell me, and the only relief I have at such times is to arrest myself, and believe that you will know the forgetfulness is _involuntary_. After all, though your science is much to me, we are not friends for science sake only, but for something better in a man, something more important in his nature, affection, kindness, good feeling, moral worth; and so, in remembrance of these, I now write to place myself in your presence, and in thought shake hands, tongues, and hearts together.

'We are all pretty well here. We get on well enough, in a manner, and are very happy, and I cannot wish you better things; though I have no intention, when I say that, to imagine you without your memory or your science. Long may you be privileged to use them, for the good of human nature.

'I have several times gone into a place of rest, to look at a stone you know of, and think of you all. Such places draw my thoughts much now, and have for years had great interest for me; they are not to me mere places of the dead, but full of the greatest hope that is set before man, even in the very zenith of his physical power and mental force. But perhaps I disturb you in calling your loss to mind; forgive me. Yet remember me very kindly to the mother and sisters.

'Ever, my dear Schönbein, yours affectionately,

'M. FARADAY.'

FARADAY TO DR. BECKER.

'The Green, Hampton Court: October 25, 1860.

'My dear Dr. Becker,—It was a great delight to me to receive your very pleasant and affectionate letter last month.

.   .   .   .   .   .

'I have been greatly interested in reading your account of your proceedings at Bonn, Heidelberg, and Giessen. I am not competent to form an opinion of the best mode of pursuing science in Germany by a

German mind; but the advice of Buff is that which
would soonest fall in with my own thoughts and ways.
I could not imagine much progress by reading only,
without the experimental facts and trials which could
be suggested by the reading. I was never able to
make a fact my own without seeing it; and the de-
scriptions of the best works altogether failed to convey
to my mind such a knowledge of things as to allow
myself to form a judgment upon them. It was so with
new things. If Grove, or Wheatstone, or Gassiot, or
any other, told me a new fact, and wanted my opinion,
either of its value, or the cause or the evidence it
could give on any subject, I never could say anything
until I had seen the fact. For the same reason, I never
could work, as some professors do most extensively,
by students or pupils. All the work had to be my
own. I know very well that my mind is peculiarly
constituted, that it is deficient in appreciation, and,
further, that the difficulty is made greater by a failing
memory. From this, nevertheless, you will understand
how my thoughts fall in with Buff's opinion, and how
terrified I should be to set about learning science from
books only. However, what we call accident has in
my life had much to do with the matter, for I had to
work and prepare for others before I had earned the
privilege of working for myself, and I have no doubt
that was my great instruction and introduction into
physical science.

.   .   .   .   .   .

' Believe me to be, my dear Dr. Becker, ever faith
fully yours,

' M. FARADAY.'

Mr. Stroud wrote to ask him regarding the truth of some statements[1] made by a lecturer on Paddington Green. In his answer he said :—

FARADAY TO MR. STROUD.

'The Green, Hampton Court : July 6, 1860.

' Sir,—Your letter has surprised me a good deal, for I did not know before that my name had been used as you describe, and cannot now imagine how it has been employed upon that side of the argument where your letter places it. I send herewith a part of your letter (*which, however, I will thank you to return to me again*). All that part which is between my initials on pp. 5 and 6 is utterly untrue. I never made animalcules or maggots by the agency of electricity, and when others said they had done anything of the kind, opposed their views, and all the conclusions derived from them. I never lectured on science at Cambridge at all ; no lectures of mine have been discontinued, and if I have given offence (which I can only imagine in

[1] 'When combating the Old Testament narrative of the creation of man, he (Mr. Wild) adverted to certain chemical experiments which he has alleged were made by you some years since, before audiences, both at Oxford and Cambridge, and also in London, when you demonstrated that life was but electricity, by producing through its agency animalcules, maggots, &c , accompanying those experiments by the remarks addressed to your audiences, as : "Gentlemen, there is life, and, for aught I can tell, man was so created." Mr. Wild has always held it (and has related the circumstance to show) that you inferred from your experiments that man could be created or generated, and in all probability was created, in the same *modus operandi* as by your experiments.

' Mr. W., in relating the above, has always added that so unpalatable were your views, and contrary to what was received as orthodox, that the authorities under whose auspices the lectures were given (at which you experimented) had them discontinued.'

the case of one person[1]), it has been because I was
supposed to pay too much respect to the Bible, which
I believe to be the *Word of God.*

'Some years ago I delivered a *lecture on education*,
which has since been reprinted at the end of a volume
of Juvenile Lectures on the forces of matter, just published by Griffin, I believe. Near the beginning of that
lecture you will find a public answer to the inquiries
which you make at the close of your note.

'Your letter states that the object of the meetings on
Paddington Green is the elucidation of truth. As far as
your letter goes, they appear to me to have been effectual
mainly in the generation and propagation of *error.*

'You are at perfect liberty to use this letter in
connexion with the subject in any way you may think fit.

'I am, Sir, your very faithful servant,

'M. FARADAY.'

The work in 1861 was only for the Trinity House
and for the Royal Institution.

He gave ten reports to the Trinity House. One of these
was on a visit, October 31, to Dungeness, to see the new
magneto-electric lamps, the machines, and the steam-
engines. He was also much occupied this year and the
next with the adjustment of the illuminating apparatus
to the lamp flames in the lighthouses. He devised an
apparatus for determining the course and focus of the
rays, and rendered the system of adjustment perfect.

---

[1] This was in the lecture on mental education in 1854 by the sentence
'Let no one suppose for a moment that the self-education I am about to
commend in respect of the things of this life extends to any considerations
of the hope set before us, as if man by reasoning could find out God.'

For the Institution he gave Friday evening discourses on platinum, and on Warren de la Rue's photographic eclipse results.

In October he wrote to the managers of the Institution :—'It is with the deepest feeling that I address you. I entered the Royal Institution in March 1813, nearly forty-nine years ago, and, with exception of a comparatively short period, during which I was abroad on the Continent with Sir H. Davy, have been with you ever since During that time I have been most happy in your kindness, and in the fostering care which the Royal Institution has bestowed upon me. Thank God, first, for all his gifts. I have next to thank you and your predecessors for the unswerving encouragement and support which you have given me during that period. My life has been a happy one, and all I desired. During its progress I have tried to make a fitting return for it to the Royal Institution, and through it to science. But the progress of years (now amounting in number to threescore and ten) having brought forth first the period of development, and then that of maturity, have ultimately produced for me that of gentle decay. This has taken place in such a manner as to make the evening of life a blessing ; for whilst increasing physical weakness occurs, a full share of health free from pain is granted with it ; and whilst memory and certain other faculties of the mind diminish, my good spirits and cheerfulness do not diminish with them.

' Still I am not able to do as I have done. I am not competent to perform, as I wish, the delightful duty of teaching in the Theatre of the Royal Institution, and I now ask you (in consideration for me) to

accept my resignation of the Juvenile Lectures. Being
unwilling to give up what has always been so kindly
received and so pleasant to myself, I have tried the
faculties essential for their delivery, and I know that I
ought to retreat ; for the attempt to realise (in those
trials) the necessary points, brings with it weariness,
giddiness, fear of failure, and the full conviction that
it is time to retire ; I desire therefore to lay down this
duty. I may truly say that such has been the pleasure
of the occupation to me, that my regret must be
greater than yours need or can be.

  ' And this reminds me that I ought to place in your
hands the whole of my occupation. It is no doubt
true that the Juvenile Lectures, not being included in
my engagement as professor, were, when delivered by
me, undertaken as an extra duty, and remunerated by
an extra payment. The duty of research, superintendence of the house, and of other services, still remains ;
but I may well believe that the natural change which
incapacitates me from lecturing, may also make me
unfit for some of these. In such respects, however, I
will leave you to judge, and to say whether it is your
wish that I should still remain as part of the Royal
Institution. I am, Gentlemen, with all my heart, your
faithful and devoted servant,
                                    ' M. FARADAY.'

  Shortly afterwards he wrote to the Secretary :—' You
know my feelings, in regard to the exceedingly kind
manner in which the Board of Managers received my
letter, and *you* therefore can best convey to them my
deep thanks on this occasion. Please do this for me.
Nothing would make me happier in the things of this

1861.
Æt. 70.

life than to make some scientific discovery or development, and by that to justify the Board in their desire
to retain me in my position here.'

To the President of the Royal Institution, who
wished to resign his office, Faraday thus wrote :—

FARADAY TO HIS GRACE THE DUKE OF NORTHUM-
BERLAND.

'Royal Institution : November 15, 1861.

'My Lord Duke,—Is it essential to *yourself* that we
should lose you? You are kind, you bear with us,
you do not disturb our management, you justify it
when submitted to you, you do all that we desire. No
one can be to us the President that you are.

'Mr. Pole has shown me your Grace's note ; it, and
the remembrance of all your former kindness, makes
me thus bold to write.

'That your Grace may know of my sincerity, I quote
a small thing personal to myself. I am above seventy
years of age, and with a bad memory, feel the thought
of the Juvenile Lectures a burden to me. I have re-
tired from them ; but as the managers believe that the
remembrance of past times, and the association of my
name, is good for the Royal Institution, I still continue
engaged to do what I can for the cause of science
there. May I hope that your Grace will in the same
way continue our President.

'I should be deeply grieved if I did not think that
your Grace will forgive me the freedom of this letter.

'I am, my Lord Duke, your truly humble servant,

'M. FARADAY.'

He subscribed most liberally to the fund for raising

a monument to Sir H. Davy at Penzance. He was 1861.
made an Honorary Member of the Medical Society of Æт.69-70.
Edinburgh.

Sir Emerson Tennant wished Faraday to witness the
phenomena produced by Mr. Home. Faraday says,
in his reply : ' You will see that I consent to all this
with much reserve, and only for your sake.' Three
days afterwards, Sir E. Tennant says : ' As Mr. Home's
wife is dying, the probability is that the meeting, at
which I wished you to be present, on the 24th, may
not take place. From the same cause I am unable to
see Mr. Home previously, or to make the inquiries of
himself necessary to satisfy the queries in your letter.'

To Professor Schönbein he writes :—' You really
startle me with your independent antozone . . . .
Surely, you must hold it in your hand like a little
struggler ; for if I understand you rightly, it must be
a far more abundant body than cæsium. For the hold
you have already obtained over it, I congratulate you,
as I would do if you had obtained a crown, and more
than for a new metal. But, surely, these wonderful
conditions of existence cannot be confined to oxygen
alone. I am waiting to hear that you have discovered
like parallel states with iodine, or bromine, or hydrogen, and nitrogen. What of nitrogen? is not its
apparent quiet simplicity of action all a sham? not a
sham, indeed ; but still not the only state in which it
can exist. If the compounds which a body can form
show something of the state and powers it may have
when isolated, then what should nitrogen be in its
separate state? You see I do not work; I cannot;
but I fancy, and stuff my letters with such fancies (not
a fit return) to you.'

In another letter he says : ' I am still dull, stupefied, and forgetful.   I wish a discovery would turn up with me, that I might answer you in a decent, respectable way.   But it will not.'

Still later he says: ' I look forward to your new results with great interest ; but I am becoming more and more timid when I strive to collate hypotheses relating to the chemical constitution of matter.   I cannot help thinking sometimes whether there is not some state or condition of which our present notions give us very little idea, and which yet would reveal to us a flood, a world of real knowledge,—a world of facts available both by practical application and their illustrations of first principles : and yet I cannot shape the idea into a definite form, or reach it by any trial facts that I can devise ; and that being the case, I drop the attempt, and imagine that all the preceding thought has just been a dreaminess and no more ; and so there is an end of it.'

Writing on the subject of an angry discussion that had occurred in a family, he said :—

'August, 1861.

' We may well regret such incidents.   It is not that they are not to be expected, for they belong to our nature, but they ought to be repented.   They are exceedingly unwholesome in a moral point of view, for they generally lead each one in private to justify themselves, and so foster a pharasaical condition of mind, and a growing tendency to judge others rather than ourselves.   I speak from my own conviction.   I know that the real root in such cases is not worth a thought ; but, at the same time, I know that a vast mass of the un-

comfort of life depends upon the tendency to criticise
rather than to excuse or commend things which in
one view or another deserve the latter as much as the
former. We may well remember Hamlet, " If we all
had our deserts, who would escape a whipping?" '

FARADAY TO MISS MOORE.

'The Green, Hampton Court : August 14, 1861.

' My dear Friend,—I have been writing to you (in
imagination) during a full week, and the things I had
to talk about were so many that I considered I should
at last want a sheet of foolscap for the purpose ; but
as the thoughts rose they sank again, and oblivion
covers all. And so it is in most things with me ; the
past is gone, *not* to be remembered ; the future is coming,
not to be imagined or guessed at ; the present only is
shaped to my mind. But, remember, I speak only of
temporal and material things. Of higher matters, I
trust that the past, present, and future, are *one* with
me, and that the temporal things may well wait for
their future development.

.   .   .   .   .   .

' Ever your faithful friend,
' M. FARADAY.'

1862 was the last year of experimental research.
Steinheil's apparatus for producing the spectrum of
different substances gave a new method by which the
action of magnetic poles upon light could be tried. In
January he made himself familiar with the apparatus,
and then he tried the action of the great magnet on
the spectrum of chloride of sodium, chloride of barium,
chloride of strontium, and chloride of lithium.

On March 12 he writes : ' Apparatus as on last day (January 28), but only ten pairs of voltaic battery for the electro-magnet.

' The colourless gas flame ascended between the poles of the magnet, and the salts of sodium, lithium, &c., were used to give colour. A Nicol's polariser was placed just before the intense magnetic field, and an analyser at the other extreme of the apparatus. Then the electro-magnet was made, and unmade, but not the slightest trace of effect on or change in the lines in the spectrum was observed in any position of polariser or analyser.

' Two other pierced poles were adjusted at the magnet, the coloured flame established between them, and only that ray taken up by the optic apparatus which came to it along the axis of the poles, i.e. in the magnetic axis, or line of magnetic force. Then the electro-magnet was excited and rendered neutral, but not the slightest effect on the polarised or unpolarised ray was observed.'

This was the last experimental research that Faraday made. His scientific faith and his stedfastness are well seen in his work on the relation of electricity and magnetism to light. In 1834, he first transmitted a ray of polarised light directly across the course of the electric current (' Researches in Electricity,' vol. i. p. 285.) In 1845 ' he at last succeeded in magnetising and electrifying a ray of light (vol. iii. p. 2) ; and now in 1862 he ends his laboratory work with the same subject, with the same negative result, which he had obtained twenty-eight years before.

On June 20, he gave his last Friday discourse on gas-furnaces, for the Institution. The notes which he

made for this his last lecture are very touching and
very characteristic :—

'Personal explanation,—years of happiness here, but time of retirement; LOSS OF MEMORY and *physical endurance of the brain*.

'1. Causes—*hesitation and uncertainty* of the convictions which the speaker has to urge.

'2. *Inability to draw* upon the mind for the treasures of knowledge it has previously received.

'3. *Dimness*, and forgetfulness of one's former *self-standard* in respect of *right, dignity*, and *self-respect*.

'4. Strong duty of *doing justice to others*, yet inability to do so.

'*Retire.*'

In 1813, when writing on lecturers, he said: 'It may perhaps appear singular and improper, that one who is entirely unfit for such an office himself, and who does not even pretend to any of the requisites for it, should take upon him to censure and commend others.'

Yet in 1816, he began to lecture at the City Philosophical Society, and 'he showed his strong determination to execute everything he undertook in the best manner,' as his friend Magrath wrote of him, by entering himself at the same time to an evening class of lectures by Mr. B. H. Smart, on elocution. His lectures continuing, the following year he entered to another course, 'when his means were far from ample,' says Mr. Smart.

In 1823, previous to taking a part in Mr. Brande's laboratory lectures, he took private lessons of Mr. Smart, 'at the rate of half a guinea each;' and in 1825 and 1826 he took more lessons, before he gave his first course in the Theatre of the Institution, in 1827. Afterwards, Mr. Smart ' often attended his lectures, in order to provide himself with ground for remark on his address and delivery.'

The following rules were found among his notes :—

' Never to repeat a phrase.

' Never to go back to amend.

' If at a loss for a word, not to ch-ch-ch or eh-eh-eh, but to stop and wait for it. It soon comes, and the bad habits are broken, and fluency soon acquired.

' Never doubt a correction given to me by another.'

For thirty-eight years his lectures were the life of the Royal Institution. His singular power of making himself one with his audience was felt in his Juvenile Lectures, in his theatre courses, and in his Friday evening addresses.

In his Juvenile Lectures, his simple words and his beautiful experiments, his quickness and his clearness, kept the attention, and fixed his instruction in the mind even of the youngest of his hearers, whilst the most practised teacher would find old experiments shown in a new form, which the genius of Faraday only could have invented, and which his handicraft enabled him to carry out.

In his theatre lectures, his matter was always overabundant, his experiments were always successful, his knowledge was always at the furthest limits to which it had at the time been extended by himself or by others, and yet his consideration for those who knew but little never failed.

But it was in the Friday evening discourses that his great power as a lecturer came out. His manner was so natural, that the thought of any art in his lecturing never occurred to anyone. Rapidly and yet clearly, he made the object of his lecture known. Those who had but little knowledge could see his starting point, and they thought they saw where he was going. Those who knew most followed him beyond the bounds of their own knowledge, forgetting almost the lecturer, who seemed to forget himself in his words and his experiments, and who appeared to be trying only to enable them to judge what his latest discoveries were worth ; and when he brought the discoveries of others before his hearers, one object, and one alone, seemed to determine all he said and did, and that was, 'without commendation and without censure,' to do the utmost that could be done for the discoverer.

For the Juvenile Lectures, Faraday had always received 50*l.* This year the Institution added this sum to the 400*l.* which were paid to him as Director of the House and Laboratory, and as Fullerian Professor of Chemistry.

He gave seventeen reports to the Trinity House and two to the Board of Trade. The most important of the Trinity House reports were still on the magneto-electric light.

On February 12, he went to observe the new light at Dungeness and to examine the keepers. He slept at the lighthouse, and was joined by the Deputy Master in the yacht. At night he went to sea, testing at five miles off the effects of the oil-lamp, reflectors, and the electric light, Professor Holmes himself being in charge of the lamps for the trials. Afterwards, he

went to the Varne floating light, and compared Dungeness, Grisnez, and the South Foreland lights. He
slept at Dover, and after examining the upper South
Foreland new hydrostatic lamp, he returned home.
On May 12 and 13 he was again at Dungeness with
Mr. Chance, and came to the conclusion 'that the
present apparatus is abundantly sufficient to supply
every proof that can be desired to establish the fitness
(or the contrary), of the magneto-electric light for
lighthouse purposes.'

Observations, according to a form drawn up by
Faraday, were kept at Upper and Lower South Foreland lighthouses, at the Varne, and by four pilot
cutters; and a special observer, trained by Faraday to
make measurements of the intensity of the light, was
sent for six months to the Varne.

This year he was examined at great length by the
Public School Commissioners. His most memorable
answers were these: 'That the natural knowledge
which had been given to the world in such abundance
during the last fifty years, I may say, should remain
untouched, and that no sufficient attempt should be
made to convey it to the young mind, growing up and
obtaining its first views of these things, is to me a
matter so strange that I find it difficult to understand;
though I think I see the opposition breaking away, it
is yet a very hard one to be overcome. That it ought
to be overcome, I have not the least doubt in the
world.' In answer to the question at what age it
might be serviceable to introduce the physical sciences,
he says: 'I think one can hardly tell that until after
experience for some few years. All I can say is this,
that at my Juvenile Lectures at Christmas time, I have

never found a child too young to understand intel- 1862.
ligently what I told him; they came to me afterwards Æt.70–71.
with questions which proved their capability.'

Again he says: 'I do think that the study of
natural science is so glorious a school for the mind,
that with the laws impressed on all created things by
the Creator, and the wonderful unity and stability of
matter and the forces of matter, there cannot be a
better school for the education of the mind.'

The Duke of Devonshire, at his installation as
Chancellor, was anxious that the University of Cambridge should confer the degree of LL.D. on Faraday,
' as a practical proof would be afforded of the University's having made some advance in liberality of sentiment.' Thus, under the new statutes, which rendered
a profession of belief in the Thirty-nine Articles unnecessary, almost the last (as well as the first) honour
he received came from Cambridge. He was also made
Knight Commander of the Order of St. Maurice and
Lazarus, Italy.

Two letters, very characteristic of his affection and
resignation, were written this year, the one to Mrs.
Faraday, the other to Mrs. Barlow; and a very sad
note to Professor Schönbein brought their long correspondence to an end. ' He was so changed, *I could*
not write any more,' said Professor Schönbein to me.

FARADAY TO MRS. FARADAY.

'Brighton: Monday, September 15,
(December 15 post mark) 1862.

' Home safely, dearest. My heart is with you all.

' Ever yours,
' M. FARADAY.'

To Mrs. Barlow he said :—

'I write you a few words of sympathy and remembrance. I am not well enough to write many. My throat is very sore, and age tells with the attack, but I know you would wish a few words rather than none. I called at your house, and I rejoice to think that your absence is a sign of good health. *Our love to you both.* I am enjoying the gradual decay of strength and life, for when I revive it is no great revival or desire to me, and that cheers me in the view of death near and round us. We think it impossible that my dear sister can last much longer.'

In September he wrote his last letter to Professor Schönbein. He says :—

'Again and again I tear up my letters, for I write nonsense. I cannot spell or write a line continuously. Whether I shall recover — this confusion — do not know. I will not write any more. My love to you.

In 1863 the most important work that was done was for the Trinity House. He made twelve reports, and among them was his final report upon the magneto-electric light as applied to lighthouses.

On February 17 and 18, he was again at Dungeness. 'The examination was in every point satisfactory.' The chief object was to see a new optic apparatus for the magneto-electric light ; the examination on the night of the 18th, from the Varne, fully confirmed the expectations raised by the examination of the previous day. Everywhere 'the new apparatus sustained its superiority over the old.'

In June, at the end of his report on the monthly
observations made upon the electric light, he says :
'The other reports require no especial comment,
but I think the general character and action of the
magneto-electric light at Dungeness is now so well
established by the accumulated evidence from these
various outlying places of observation, that their continuance is hardly required, and that it is only the
general surveillance, with the record of any special
failure or departure from the ordinary result, that will
be required in future. No tabular form can be given
for these ; they must depend upon the watchfulness and
willingness of the parties having the opportunities.'

In September, when making a report upon a communication to the Trinity House from the Board of
Trade regarding Berlioz's magneto-electric machine,
which Faraday had examined at the Exhibition of 1862,
he says :—

'That the magneto-electric machine is admirably
adapted for application in lighthouses is not a conclusion to be drawn from principle, but one that has
been thoroughly established by practice at the South
Foreland and Dungeness lighthouses, i.e. so long as the
knowledge, attention, and precautions are secured that
are necessary for the uninterrupted command and
security of the light.'

Thus then, the application of the magneto-electric
light to lighthouses by Professor Holmes, passed
through the second stage of its probation.

After examining the results with the most watchful
care, and the most unbiassed judgment, Faraday was
able to recommend that his own grandest discovery

should be applied to 'the great object of guiding the mariner across the dark and dreary waste of water.'

He was made Foreign Associate of the Imperial Academy of Medicine, Paris, and of the National Academy of Sciences of the United States.

The following letters to his wife, and his niece, Miss Reid, abound in evidence of his love for his home.

FARADAY TO MRS. FARADAY.

'Dungeness, Wednesday: February 13, 1863.

'Dearest,—Here at the lighthouse at ten o'clock P.M., all successful so far, but when you will get this I do not know. I shall put it in the post to-morrow. It is now bedtime, and as we have made our observations here pretty well, I shall be able to start to-morrow for Dover. The weather is very good for us here, and I hope to be home on Friday; perhaps in the middle of the day. Remember me; I think as much of you as is good for either you or me. We cannot well do without each other. But we love with a strong hope of love continuing ever, in which hope my dear Jeannie joins us. My love to you both,

'Ever yours,

'M. FARADAY.'

Another letter to his wife may well be compared in its affection with one of the letters which he wrote from Swansea, forty-one years previously.

FARADAY TO MRS. FARADAY.

'5, Claremont Gardens, Glasgow: Monday, August 14, 1863.

' Dearest,—Here is the fortnight complete since I left you, and the thoughts of my return to *our home* crowd in strongly upon my mind. Not that we are in any way uncared for, or left by our dear friends, save as I may desire for our own retirement. Everybody has overflowed with kindness, but you know their manner, and their desire, by your own experience with me.

' I long to see you, dearest, and to talk over things together, and call to mind all the kindness I have received. My head is full, and my heart also, but my recollection rapidly fails, even as regards the friends that are in the room with me. You will have to resume your old function of being a pillow to my mind, and a rest, a happy-making wife.

' My love to my dear Mary. I expect to find you together, but do not assume to know how things may be.

' Jeannie's love with mine, and also Charlotte's, and a great many others which I cannot call to mind.

' Dearest, I long to see and be with you, whether together or separate

' Your husband, very affectionate,

' M. FARADAY.'

FARADAY TO HIS NIECE, MISS REID.

' The Green, Hampton Court: October 1, 1863.

' Dear Maggie,—So we turn the times over. Here is the first of a new month, and a new season is coming over us, for the rain falls, and the winds blow, and the sun shines with a strength and in an order, or rather

disorder, that reminds me of an old man, who, pur-
posing to do one thing is drawn off to do another, who,
intending to communicate with you, is led away to chat
with another girl, and forgets you. However, she is
gone. I forget what the thought was like ; its end is confusion, and so I come to wakefulness and life again.'

Then he writes regarding the serious illness of two
relations, and continues: ' All these events may well
lessen our thoughts and hold of life. But what a
blessing it is that there is nothing in them to diminish
the hopes belonging to that far better life to which
this is only the entrance.

' The sorrow is for the night only ; *joy cometh in the
morning.*

' Give my love to your dear father.

' You know I am clumsy at sending loves, and wish
not to be mingled with those who do it as of habit,
but here are a few names besides those I have mentioned.

.   .   .   .   .   .

' I must stop, or I shall run into folly.

' Ever, my dear Maggie, your affectionate uncle,

' M. FARADAY.'

FARADAY TO MISS MOORE.

' The Green, Hampton Court: September 17, 1863.

' My dear Friend,—Many thanks for your lightning-
like letter ; like and unlike—for it was rather slow in
its progress, having been mis-sent to Southampton, vide
enclosure ; and yet, in that, like the lightning, which
often falls in very unexpected places.

' Lightning is a very curious thing ; I have often seen
the course of the discharge upon trees, beginning

suddenly, and ending as unexpectedly as some of those your brother speaks of. We have to remember that the electricity is not always as a vivid, concentrated, explosive flash, throughout the whole of its course. The cloud, or the air over a tree, being highly charged, may induce torrents upon it, but the first progress of the electricity may be, in fact, invisible streams or brushes, which, as they come together, accumulate and break out into one luminous, concentrated, and powerful spark. We can easily produce an effect of this kind by our ordinary electrical machines, when working upon the conversion of sparks and brushes into each other. I have several times seen trees, which, having been struck by lightning, have exhibited afterwards the beginning and the ending of the visible barked place : the beginning having occurred at the angle where one branch separated from another : and the ending or bottom at a larger part of the trunk, lower down. By examining the branches carefully upwards, I have seen reason to believe—1st, that when the atmospheric electricity first took its course to and through the tree, it has fallen on the leaves and fine stems, chiefly as brushes, or in the non-luminous and brush state. 2nd, that as these have been conducted downwards, they have run together, and made more concentrated streams ; that the concentration has resulted in the production of one powerful, luminous, heating, and explosive discharge. 3rd, that as the quantity of electricity was continually diminished by the conductive force of the tree (as the electricity neared the ground, and the mass of the trunk became a larger, and therefore a more effective conductor at the lower parts), it might well be that a discharge of electricity (appearing as a bright flash in one part of the

trunk), would before it reached the bottom be altogether *conducted*; and then would lose its luminous
character.

'As to the difference between the edges and the
middle of the ruptured place, I have not seen the case,
therefore have no right to form an opinion; but what
say you to these thoughts? the tear occurs in a solid
resisting body; it is perhaps an inch wide, and several
feet long; the dispersive force is at right angles to the
length of the mark. Can we not imagine that the
escape of the disturbed particles is easier on the two
sides or edges of the course of the explosion, than at
the middle, and can that circumstance conduce to the
difference?

'Ever, my dear friend, yours truly,
'M. FARADAY.'

At the request of the managers of the Royal Institution, Faraday asked the Prince of Wales to become
Vice-Patron.

FARADAY TO HIS ROYAL HIGHNESS THE PRINCE OF WALES.

'Royal Institution: January 5, 1863.

'May it please your Royal Highness,—The President,
Secretary, and Managers, remembering the great grace
which was done to the body when the late Vice-
Patron, His Royal Highness the Prince Consort, with
yourself and the Prince your brother favoured our
lectures with your presence, have requested me to
express the hope they entertain that your Royal Highness will allow of your election into the position
then filled by your admirable and much-loved father.
It is because I had the honour of speaking at that time

in your presence, that the authorities here think I may
not be unacceptable as the channel through which
their earnest desire may be made known. Our most
honoured Queen is our Patron ; and that we may
have your Royal Highness as her and our Vice-Patron
is the most earnest wish of our hearts. Remembering
the past, we hope for this great grace for the future.
With the deepest feelings of respect, I venture to sign
myself,

'Your Royal Highness's faithful servant,

'M. FARADAY.'

On receiving a volume of the Prince Consort's
speeches from the Queen, he thus wrote to Sir James
Clark :—

FARADAY TO SIR JAMES CLARK.

'Royal Institution : February 7, 1863.

'My dear Sir James,—Her Majesty our Queen has
done me great honour (and a favour most especially
welcome) in thinking of me in relation to our most
worthy and glorious Prince, his late Royal Highness the
Prince Consort. I do not know how to thank Her
Majesty enough or well—may I hope that you will
help me ? I would, if I might, express my reverence
for the Queen, the wife and the mother whose image
dwells in the hearts of all her people. I wish that I
were, as a subject, more worthy of her ; but the vessel
wears out, and at seventy-one has but little promise for
the future. The fifty years of use in the Royal Institution has given me wonderful advantages in learning,
many friends, and many opportunities of making my
gratitude known to them ; but they have taken the
matter of life, and above all, memory out of me, leaving

the mere residue of the man that has been, and now I remain in the house useless as to further exertion, excused from all duty, very content and happy in my mind, clothed with kindness by all, and honoured by my Queen.

'Ever, my dear Sir James, your most faithful servant,
'M. FARADAY.'

FARADAY TO DR. CARPENTER,

THE REGISTRAR OF THE LONDON UNIVERSITY.

'Royal Institution : April 20, 1863.

'My dear Dr. Carpenter,— Many of your recent summons have brought so vividly to my mind the progress of time in taking from me the power of obeying their call, that I have at last resolved to ask you to lay before the senate my desire to relinquish my station and render up that trust of duty which I can no longer perform with satisfaction, either to myself or to others.

'The position of a senator is one that should not be held by an inactive man to the exclusion of an active one. It has rejoiced my heart to see the progress of the university, and of education under its influence and power ; and that delight I hope to have so long as life shall be spared to me.

'Ever, my dear Dr. Carpenter, yours most truly,
'M. FARADAY.'

FARADAY TO DR. HOLZMANN.

'Royal Institution : December 22, 1863.

'My dear Sir,—I have just risen (at midday) from my bed to acknowledge your very pleasant and kindly

letter. My words totter, my memory totters, and now
my legs have taken to tottering, and I am altogether a
very tottering and helpless thing.

'I rejoice to hear that the Queen looks on us with
such favour as to decide that His Royal Highness Prince
Arthur should attend some of the lectures. You may
be sure of my earnestness in the cause, and you cannot
think how much Her Majesty's sanction and approval
of our course of action stimulates all here to persevere
in the course which has thus far gained the approbation of Her Majesty, and our beloved member the late
Prince Consort.

'Ever, my dear Dr. Holzmann, yours most truly,
'M. FARADAY.'

In 1864, twelve reports were made between January
and October to the Trinity House. One was on a new
magneto-electric machine; another on drawings, pro-
posals and estimates for the magneto-electric light at
Portland. He made seven examinations of white and
red leads, and two examinations of waters from Orfordness and the Fog-gun station, Lundy Island; and he
reported on two 4th-order lights for the River Gambia.

At the end of the year he was asked by Mr. Cole to
be a Vice-President of the Albert Hall. He replies:—
'I have just returned from Brighton, to which place
my doctor had sent me under nursing care. Hence
the delay in answering your letter, for I was unaware
of it until my return. Now, as to my acceptance of
the honour you propose to me. With my rapidly
failing faculties, ought I to accept it? You shall
decide. Remember that I was obliged to decline
lecturing before Her Majesty and the Royal Family at

Osborne ; that I have declined and am declining the
Presidency of the Royal Society, the Royal Institution,
and other bodies; declaring myself unfit to undertake
any responsibility or duty even in the smallest degree.
Would it not therefore be inconsistent to allow my
name to appear amongst those of the effectual men
who delight, as I should have done under other
circumstances, to honour in every way the memory of
our most gracious and regretted leader ? These are my
difficulties.   It is only the name and the remembrance
of His Royal Highness which would have moved me
from a long-taken resolution.'

Mr. Cole decided, 'without a moment's doubt, that
he was to be a Vice-President.'

He was made Foreign Associate of the Royal Academy
of Sciences, Naples.

A letter which he wrote this year to his friend Miss
Moore shows how the feeling of increasing weakness
was growing upon him : ' I find myself less and less fit
for communication with society ; even in a meeting of
family—brothers and sisters—I cannot keep pace in
recollection with the conversation, and so have to sit
silent and taciturn.   Feeling this condition of things, I
keep myself out of the way of making an exposure of
myself.   Remaining life is only for my friends who
receive me for past affection's sake, and also for what I
am. Now I count you amongst these, so notwithstanding
bad spelling and other deficiencies, keep up the public
current of kind communication.'

He wrote another letter to a niece which gives a
vivid picture of that wonderful humility which was one
of the great characteristics of his mind :—

FARADAY TO A NIECE.

'Royal Institution : April 10, 1864.

' My very dear Niece and Friend,—This is the evening of the Sabbath day. We have just come home, bringing . . . . with us ; have had our tea, and by this most agreeable though waning light, I propose to thank you for your kind and acceptable note. . . .

' I am at present as well as I think any man at my age has any reason to expect to be, and in many points I am much better. It is true my memory is much gone, nearly all gone ; and the power of recollection is nearly lost, *as to precision.* But then all about me are very kind. My worldly friends remember the times past, and do not want me to give up my posts or pay, yet willingly remit the work ; and then He who rules over all is kinder than all ; and though I sometimes tremble when I have occasion in doctrine or judgment to use His word, being unable to remember it, I dare not venture to put that (his Eldership) from me which He has put upon me ; and I call to mind that His throne is a throne of grace, where prayer may be made for help and strength in time of need. And He makes my brethren so kind, that there is only one of the number who teases me, and that is myself, and I often think pride and the absence of humility has much to do with that. . . . . . '

He thus replied to an invitation from the Messrs. Davenport :—

FARADAY TO MESSRS. DAVENPORT.

'October 3, 1864.

'Gentlemen,—I am obliged by your courteous invitation, but really I have been so disappointed by the manifestations to which my notice has at different times been called, that I am not encouraged to give any more attention to them, and therefore I leave these to which you refer in the hands of the professors of legerdemain. If spirit communications not utterly worthless, of any worthy character, should happen to start into activity, I will leave the spirits to find out for themselves how they can move my attention. I am tired of them.

'With thanks, I am very truly yours,

'M. FARADAY.'

To a gentleman who invited him to some spiritual manifestations he sent these answers :—

FARADAY TO THOS. S———, ESQ.

'Royal Institution; November 1, 1864.

'Sir,—I beg to thank you for your papers, but have wasted more thought and time on so-called spiritual manifestation than it has deserved. Unless the spirits are utterly contemptible, *they* will find means to draw my attention.

'How is it that your name is not signed to the testimony that you give? Are you doubtful even whilst you publish? I've no evidence that any natural or unnatural power is concerned in the phenomena that requires investigation or deserves it. If I could consult the spirits, or move them to make themselves honestly manifest, I would do it. But I cannot, and am weary of them.

1864.

Æt. 73.

'I am, Sir, your obedient servant,

M. FARADAY.'

TO THE SAME.

'Royal Institution : November 4, 1864.

'Sir,—I beg to acknowledge your letter of the 3rd, but I am weary of the spirits—all hope of any useful result from investigation is gone ; but as some persons still believe in them, and I continually receive letters, I must bring these communications to a close. Whenever the spirits can counteract gravity or originate motion, or supply an action due to natural physical force, or counteract any such action ; whenever they can punch or prick me, or affect my sense of feeling or any other sense, or in any other way act on me without my waiting on them ; or working in the light can show me a hand, either writing or not, or in any way make themselves visibly manifest to me ; whenever these things are done or anything which a conjuror cannot do better ; or, rising to higher proofs, whenever the spirits describe their own nature, and like honest spirits say what they can do, or pretending, as their supporters do, that they can act on ordinary matter whenever they initiate action, and so make *themselves* manifest ; whenever by such-like signs they come to me, and ask my attention to them, I will give it. But until some of these things be done, I have no more time to spare for them or their believers, or for correspondence about them.

'I am, Sir, yours very truly,

'M. FARADAY.'

FARADAY TO REV. W H. M. CHRISTIE

(ON THE DEATH OF HIS FATHER).

'Royal Institution: January 25, 1865.

'My dear Sir,—Very many thanks for your note. I am not surprised by the information, but greatly interested. For I am as one waiting the call and warned by each example of an old friend. I am glad he and you were spared the example of pain, but the dealings of God are very merciful.

'Do not think of me as regards the funeral. I could not attend it if I would, and such things are to me only formal. I shall always remember him whilst time is left to me as a dear old friend. I hope the family will be comforted.

'Ever yours truly,
'M. FARADAY.'

In May 1865, he made his last report for the Trinity House. It was on St. Bees light.

He wrote to the Deputy Master :—

'I write to put myself plainly before you, in respect of the matter about which I called two days ago. At the request of the then Deputy Master, I joined the Trinity House in February, 1836, now near upon thirty years since. I find that time has had its usual effect upon me, and that I have lost the power of remembering, and also of other sorts, and I desire to relieve my mind. Can this be done without my retiring altogether, and can you help me in this matter?'

Arrangements were immediately made, by which Dr. Tyndall undertook the work for Faraday.

1865.
Æт.73-74.

From first to last, his engagement with the Trinity House showed the noble spirit that was in him.

When he was appointed scientific adviser in 1836, he told the Deputy Master 'of his indifference to his proposition as a matter of interest, though *not as a matter of kindness.'*

The work which he did included the ventilation of lighthouses, the arrangements of their lightning conductors, the analysing and supervising of their drinking waters ; the examination of their optical apparatus ; the determination of the worth of the different propositions made to the Trinity House regarding the lights, extending from the practical use of the magneto-electric light, even down to the samples of cottons, oils, and paints that were to be used.

His knowledge, judgment, accuracy, and dutiful service were repaid by an appointment known only to very few persons,—an unlimited amount of kindness, and 200*l.* a year.

It will be said that the Trinity House, like the Royal Institution, was only a private company. Perhaps, like the Institution, it would gladly have given more money, but it was only able to give kindness, which to Faraday was beyond all money and all fame. Even the work which he did for the Government, he did 'for love,' and not for pay. In his letter to Lord Auckland, he said : 'I have always, as a good subject, held myself ready to assist the Government, if in my power, not for pay, for, except in one instance * (and then only for the sake of the person joined with me), I refused to take it.'

All that he did for the Government, the Royal In-

* The Haswell Colliery investigation.

stitution, and the Trinity House, was done for the honour and service of his country.

Writing to Lord Wrottesley regarding rewards for scientific men, he said : ' For its *own sake*, the Government should honour the men who do honour and service to the country.' 'I have, as a scientific man, received from foreign countries and sovereigns, honours which surpass, in my opinion, anything which it is in the power of my own to bestow.' Most noble England !

For his Trinity House work, Faraday did receive the highest reward a scientific man can obtain, but it did not come from the Government nor from the Trinity House. He was able to report that his own grandest discovery could be made useful for the preservation of the lives of seamen.

This year Sir David Brewster sent him a pamphlet on the invention and introduction of the dioptric lights, and asked him to give his opinion on the value and importance of these lights. He replied :—

' . . . . I would rather not enter as an arbitrator or judge into the matter, for I have of late been resigning all my functions as one incompetent to take up such matters, and the Royal Institution, as well as the Trinity House, have so far accepted them as to set me free from all anxiety of thought in respect to them. In fact, my memory is *gone*, and I am obliged to refrain from reading argumentative matter, or from judging of it. I am very thankful for their tenderness in the matter ; and if it please Providence to continue me a year or two in this life, I hope to bear the decree patiently. My time for contending for temporal honours is at an end, whether it be for myself or others.'

To the managers of the Royal Institution he wrote,
March 1 :—

FARADAY TO THE MANAGERS OF THE INSTITUTION.

' Unless it be that as I get older I become more infirm in mind, and consequently more timid and unsteady, and so less confident in your warm expressions, I might, I think, trust more surely in your resolution of December 2, 1861, and in the reiterated verbal assurances of your kind secretary, than I do; but I become from year to year more shaken in mind, and feel less able to take any responsibility on me. I wish, therefore, to retire from the position of superintendent of the house and laboratories. That which has in times past been my chiefest pleasure has now become a very great anxiety; and I feel a growing inability to advise on the policy of the Institution, or to be the one referred to on questions both great and small, as to the management of the house.

' In a former letter, when laying down the Juvenile Lectures, I mentioned " that other duties, such as research, superintendence of the house, and other services still remain; " but I then feared that I might be found unfit for them; I am now persuaded that this is the case. If under these circumstances you may think that with the resignation of the positions I have thus far filled the rooms I occupy should be at liberty, I trust that you will feel no difficulty in letting me leave them, for the good of the Institution is my chief desire in the whole of this action.

' Permit me to sign myself personally, your dear, indebted, and grateful friend,   ' M. FARADAY.'

1865.
Ær.73–74. At the meeting of the managers it was resolved unanimously—

'That the managers thank Professor Faraday for the scrupulous anxiety which he has now and ever shown to act in every respect for the good of the Royal Institution. They are most unwilling that he should feel that the cares of the laboratories and the house weigh upon him. They beg that he will undertake only so much of the care of the house as may be agreeable to himself; and that whilst relinquishing the duties of " Director of the laboratory," he will retain his home at the Royal Institution.'

Before this resolution was proposed, the Secretary wrote to Faraday to ask whether it could be made more in accordance with his wishes. He answered :—

FARADAY TO THE SECRETARY OF THE INSTITUTION.

'My dear Dr. Bence Jones,—I would not have a word altered of that which marks so truly the kindness of the managers and yourself, and I would say as little as possible to hide in any way that kindness from myself. Believe me that I thank you for the great continuance of it: and thank them from me. I still live in hopes that I may occasionally *deserve* the continuance of it, though I know not how.

'Ever, most truly yours,

'M. FARADAY.'

At the annual meeting of the members in May, a resolution was passed expressing their sympathy with ' their valued and esteemed friend in his indisposition.'

He signed the following note to the Secretary :—

FARADAY TO THE SECRETARY OF THE INSTITUTION.

'Hampton Court Green: May 5.

' My dear Friend,—I have received through Mr. Vincent a copy of the resolution passed last Tuesday concerning me ; I feel much gratified by their warm remembrance, and I now write to ask you to return my affectionate thanks at the next opportunity.

'I am, dear Dr. Bence Jones, your friend affectionately,

' M. FARADAY.'

To Sir James South, who wished to have some account of Anderson's services, Faraday wrote :—

' Whilst endeavouring to fulfil your wishes in relation to my old companion, Mr. Anderson, I think I cannot do better than accompany some notes which he has himself drawn up and had printed, by some remarks of mine, which will show how, and how long, he has been engaged here.

' He came to assist in the glass-house for the service of science, in September 1827, where he remained working until about 1830. Then for a while he was retained by myself. In 1832 he was in the service of the Royal Institution, and paid by it. From that time to the present he has remained with that body, and has obtained their constant approbation. In January, 1842, they raised his pay to 100*l.* per annum, with praise. In 1847 they raised it in like manner to 110*l.* For the same reason, in 1853, they raised it to 120*l.* ; and in 1860, in a minute, of which I think Mr. Anderson has no copy, they say that, in consideration of his now lengthened services, and the diligence exhibited by him, they are of opinion that his salary should be raised to 130*l*.

'Mr. Anderson still remains with us, and is in character what he has ever been. He and I are companions in years, and in work, and in the Royal Institution. Mr. Brande's testimony, when he left the Institution, is to the same purport as the others. Mr. Anderson was seventy-five years of age on the 12th of last month (January). He is a widower, but has a daughter keeping his house for him. We wish him not to come to the Royal Institution, save when he is well enough to make it a pleasure; but he seems to be happy being so employed.'

He showed the state of his health and also his anxiety to do all he could in an answer which he sent to one of the Ministers who wrote to him regarding the Cattle Plague. 'I would gladly be of use if I can, but I fear it. Your Lordship knows that I resigned the Senate of the University of London, because loss of faculty, and especially memory, took away any useful power, and that loss goes on increasing. Yesterday was my seventy-fourth birthday, and that does not promise any improvement. Nor have I any medical education or experience to give me the force necessary. I would gladly help if I could, and have often thought of applying to Miss B. Coutts, but have felt discouraged. Does your Grace think it is desirable under such circumstances to place me on such a commission?

'M. FARADAY.'

In the fine summer at Hampton Court he sat in his window delighting in the clouds and the holiday-people

on the green. A friend from London asked him how 1865.
he was. ' Just waiting,' he replied. This he had said Ær.73–74.
more fully to H.R.H. the Count of Paris in answer to
an invitation to Twickenham earlier in the year.

FARADAY TO H.R.H. THE COUNT OF PARIS.

'21 Albemarle Street: February 7, 1865.

' How shall I answer your R.H. for the handsome
and most courteous letter which I have received, seeing
it must indeed be by an apology for not accepting the
grace offered me, but I think that the weakness, and
infirmities of old age, and above all the loss of memory,
which makes this necessary, will be the sufficient and
acceptable reason.

' I bow before Him who is Lord of all, and hope
to be kept waiting patiently for His time and mode of
releasing me according to His Divine Word, and the
great and precious promises whereby His people are
made partakers of the Divine nature.

' My deepest thanks are due to Her Royal Highness
the Princess. Accept me with my weaknesses, and
believe me to be your Royal Highness's grateful and
humble servant,

' M. FARADAY.'

In January 1866, Anderson died, and Sir James
South wished some monument to be put up to him,
and wrote to Faraday. He replied :—' My dear old
friend, I would fain write to you, but, indeed, write to
no one, and have now a burn on the fingers of my
right hand which adds to my trouble ; so that I still
use my dear J.'s hand as one better than my own, and

fear I give her great work by so doing. She has, I understand, written to you this morning, and told you how averse I am to meddling with sepulchral honours in *any* case. I shall mention your good will to Anderson,' [here Faraday took the pen, because his niece made some objection to the words ' mention the good will to Anderson,' who was dead] ; ' but I tell them what are my feelings. I have told several what may be my own desire. To have a plain simple funeral, attended by none but my own relatives, followed by a gravestone of the most ordinary kind, in the simplest earthly place.

' As death draws nigh to old men or people, this world disappears, or should become of little importance. It is so with me ; but I cannot say it simply to others | here he gave up his writing, and his niece finished the note], for I cannot write it as I would.

' Yours, dear old friend, whilst permitted,

' M. FARADAY.'

During the winter he became very feeble in all muscular power, but he took the greatest interest in the description which Mr. Wilde sent him of his new magneto-electric induction machine. He wrote the date of the letter upon each page of the manuscript ; and almost the last pleasure he showed in scientific things was in a long spark given by a Holtz electric machine.

In the spring, for a short time, with decreasing power, there was occasional wandering of the mind. One day he fancied he had made some discovery somehow related to Pasteur's dextro- and lævo-racemic acid. He desired the traces of it to be carefully preserved, for ' it might be a glorious discovery.'

Mrs. Faraday had told me many months before this time, that he intended her to give me after his death his second edition of the works of Shakespear. In the summer I received the volume with the following note, which he had dictated and signed.

FARADAY TO DR. BENCE JONES.

'Hampton Court Green: July 23, 1866.

' My dear Friend,—It is my wish you should have this volume, while I am still able to have a voice in the matter. It will be a remembrance of the affection of yours, 'Ever sincerely,

'M. FARADAY.'

The Society of Arts this summer gave him the Gold Albert medal for his discoveries in chemistry, electricity, and other branches of physical science, which in their application to the industries of the world have largely promoted arts, manufactures, and commerce.

His loss of power became more and more plain during the autumn and winter : all the actions of the body were carried on with difficulty ; he was scarcely able to move ; but his mind continually overflowed with the consciousness of the affectionate care of those dearest to him.

In 1867, his niece, Miss Reid, thus writes :—

'April 22.

' This day we begin another visit to Hampton Court. This is the ninth year that we have had the privilege of seeing spring blossoms in this pleasant comfortable house by the kind invitation of my dear uncle and aunt ; they have always said (and I believe it) that their own enjoyment of the house has been heightened by the power it has given them of sharing its benefits with others. . .

'This year we came with a melancholy thought of dear uncle's declining, half-paralysed state. . . . .

'We remained three weeks, till the middle of May, and then my uncle, aunt, and Jane took up their abode there until the end came.

'I spent June at Hampton Court. Dear uncle kept up rather better than sometimes, but oh! there was always pain in seeing afresh how far the mind had faded away. Still the sweet unselfish disposition was there, winning the love of all around him.

'Very gradual had been the weaning, and the time was far past when we used to look to him on every occasion that stirred our feelings. When any new object attracted our notice, the natural thought always was, what would our uncle think of this?

'There was always something about him which particularly attracted confidence. In giving advice, he always went back to first principles, to the true right and wrong of questions, never allowing deviations from the simple straightforward path of duty to be justified by custom or precedent, and he judged himself strictly by the same rule which he laid down for others.

'I shall never look at the lightning flashes without recalling his delight in a beautiful storm. How he would stand at the window for hours watching the effects and enjoying the scene; while we knew his mind was full of lofty thoughts, sometimes of the great Creator, and sometimes of the laws by which He sees meet to govern the earth.

'I shall also always connect the sight of the hues of a brilliant sunset with him, and especially he will be present to my mind while I watch the fading of the tints into the sombre gray of night. He loved to have us with him, as he stood or sauntered on some open spot, and spoke his thoughts perhaps in the words of Gray's Elegy, which he retained in memory clearly long after many other things had faded quite away. Then, as darkness stole on, his companions would gradually turn indoors, while he was well pleased to be left to solitary communing with his own thoughts.'

.   .   .   .   .   .

One day, a few weeks before his death, one who was sitting with him at the window said, 'See, dearest, what a beautiful rainbow!' He looked at it with a happy look, and said, 'Yes, the rainbow has set its testimony in the heavens.'

On the 26th of August his niece, Miss Barnard, thus wrote to me at Spa:

'Hampton Court, August 26, 1867.

' My dear Dr. Bence Jones,—Our cares are over : our beloved one is gone. He passed away from this life quietly and peacefully yesterday afternoon. Almost immediately after you saw him, a little more than a fortnight ago, he became weaker, and has said very little to us or taken much notice of anything from that time ; but still we did not expect the change until an hour or two before it happened.

'He died in his chair, in his study ; and we feel we could desire nothing better for him than what has occurred.

'My aunt and I feel that you were the last friend that he showed any lively interest in, and we are very glad you saw him before you went away.

.   .   .   .   .   .

'I am, dear Dr. Jones, yours very sincerely,
'JANE BARNARD.'

On September 3 she again wrote to me :—

. . . . . .

'The funeral took place on Friday (the 30th), leaving here at 9·30, and taking up some of the mourners at the Royal Institution, and from thence to Highgate. By my dear uncle's verbal and written wishes, it was strictly private and plain. We could not but follow out his last wishes. I must not lead you to think we did not fully enter into his views, but some would have liked it otherwise.

. . . . . .

'My occupation has gone.

'Ever most sincerely and gratefully yours,

'JANE BARNARD.'

In conclusion, in the fewest words I will state his chief characteristics as a philosopher, and shortly put together the greatest of those qualities which made the beauty and nobleness of his character.

As a philosopher, his first great characteristic was the trust which he put in facts. He said of himself, 'In early life I was a very lively imaginative person, who could believe in the "Arabian Nights" as easily as in the "Encyclopædia," but facts were important to me, and saved me. I could trust a fact.' Over and over again he showed his love of experiments in his writings and lectures : 'Without experiment I am nothing.' 'But still try, for who knows what is possible?' 'All our theories are fixed upon uncertain data, and all of them want alteration and support from facts.' 'One thing, however, is fortunate, which is, that whatever our opinions, they do not alter nor derange the laws of nature.'

His second great characteristic was his imagination.

It rose sometimes to divination, or scientific second sight, and led him to anticipate results that he or others afterwards proved to be true.

Throughout his life his ideas of force and of matter differed from those held by others ; thereby he was led to form plans for the broadest and newest, as well as the exactest experiments. In one of his first lectures he spoke of realising ' the once absurd notion of the transmutation of the elements,' and obtaining ' the bases of the metals.'

The discoveries of Davy and Oersted led him into more connected ideas of force, and he imagined that there might be one great universal principle from which gravity, heat, light, electricity, magnetism, even life itself, might come.

He hoped to prove by experiment that there was more than a connection between the imponderable agents. He worked to find more even than a relationship, more than a common origin, for the forces of nature. He wanted to establish an actual identity among them, and in his search for the unity of all force he made all his great discoveries.

Later in life a new image of matter came into his mind. He immaterialised matter into ' centres of force,' and he materialised the directions in which matter tends to move into ' physical lines of force.' What he took from matter at its centres and gave to force he partly gave back to matter in the lines of its motion. By this he enlarged and added to the subjects which he thought naturally possible for experiment to attack, and to experiment he went to test his ideas, and though he failed to realise his imaginations, yet, by his genius, and truthfulness, and handicraft, he filled his experimental researches with new and connected facts, and thus he left to science a monument of himself which will last in all its grandeur for ages.

As a man, the beauty and the nobleness of his character was formed by very many great qualities. Among these the first and greatest was his truthfulness. His noble nature showed itself in his search for truth. He loved truth beyond all other things; and no one ever did or will search for it with more energy than he did.

His second great quality was his kindness (*agapê*). It was born in him, and by his careful culture it grew up to be the rule of his life; kindness to every one, always—in thought, in word, and in deed.

His third great quality was his energy. This was no strong effort for a short time, but a lifelong lasting strife to seek and say that which he thought was true, and to do that which he thought was kind.

Some will consider that his strong religious feeling was the prime cause of these great qualities; and there is no doubt that one of his natural qualities was greatly strengthened by his religion. It produced what may well be called his marvellous humility.

That one who had been a newspaper boy should receive, unsought, almost every honour which every republic of science throughout the world could give; that he should for many years be consulted constantly by the different departments of the Government, and other authorities, on questions regarding the good of others; that he should be sought after by the princes of his own and of other countries; and that he should be the admiration of every scientific or unscientific person who knew anything of him, was enough to have made him proud; but his religion was a living root of fresh humility, and from first to last it may be seen growing with his fame and

reaching its height with his glory, and making him
to the end of his life certainly the humblest, whilst he
was also the most energetic, the truest, and the kindest
of experimental philosophers.

To complete this picture, one word more must be
said of his religion.   His standard of duty was supernatural.   It was not founded upon any intuitive ideas
of right and wrong; nor was it fashioned upon any
outward expediencies of time and place; but it was
formed entirely on what he held to be the revelation of
the will of God in the written Word, and throughout
all his life his faith led him to endeavour to act up to
the very letter of it.

FARADAY'S TOMB IN HIGHGATE CEMETERY.

INDEX.

THE END.

LONDON : PRINTED BY
SPOTTISWOODE AND CO., NEW-STREET SQUARE
AND PARLIAMENT STREET